Engineering Ethics

W. Richard Bowen

Engineering Ethics

Outline of an Aspirational Approach

Professor W. Richard Bowen, FREng
i-NewtonWales
54 Llwyn y mor
Caswell, Swansea, SA3 4RD
United Kingdom

ISBN 978-1-84882-223-8 e-ISBN 978-1-84882-224-5

DOI 10.1007/978-1-84882-224-5

A catalogue record for this book is available from the British Library

Library of Congress Control Number: 2008938110

Cover design: eStudio Calamar S.L., Girona, Spain

Printed on acid-free paper

9 8 7 6 5 4 3 2 1

springer.com

Preface

Around the turn of the millennium, a young woman with outstanding academic achievements in science and mathematics applied to study engineering at a European university. She had chosen to study engineering particularly because of the opportunities she expected it would give her to make a contribution to the wellbeing of others. It happened that the university engineering department to which she applied had just been involved in the design of a vehicle for a world speed record attempt. When the young woman visited the university for interview this "triumph of technology" was presented as being a quintessential example of good engineering. However, though it was clear to her that the vehicle was technically ingenious, she also recognised that it was of no practical use. She concluded that she had misunderstood the nature of engineering, and still wishing to help others she changed her plans and studied medicine, at which she assuredly excelled.

This young woman's change of career was undoubtedly a specific loss for engineering. Additionally, it had a broader, tragic dimension; for her understanding of the purpose of engineering was more mature than that of the academics she encountered. Moreover, their imbalanced prioritisation of technical ingenuity over helping people is not uncommon within parts of the profession. The primary thesis of the present book is that a major challenge for engineering is to develop an aspirational ethical ethos that redresses this imbalance, and an outline of such an ethos will be presented. We particularly need to identify, promote and fulfil the *ethical opportunities* which engineering gives.

I am very grateful for the many discussions with colleagues which have benefited the development of the themes of this book, and also for the invitations to lecture on these themes to both engineers and philosophers – having to respond to audience questions is a very effective incentive to sharpen one's clarity of thought. I particularly wish to thank Iselin Eie Bowen for perceptive comments throughout the development of this work, and Espen Eie Bowen for careful reading of the final draft of the text. During the final preparation the encouragement and very rapid replies to my queries of Anthony Doyle and Simon Rees at Springer London have been much appreciated.

In his *Nichomachean Ethics*, Aristotle noted three possible dominant motivations in an individual's life: pleasure, honour and contemplation. I would suggest that he omitted one important and rarer dominant motivation: to care for others. The present book seeks to set forth the case for caring for others as a motivation in professional life. I would like to dedicate it to one for whom the care of others has been the primary motivation throughout her life: Anna Jøssang Eie.

Wales, September 2008

W. Richard Bowen
wrichardbowen@i-newtonwales.org.uk

Contents

Part I

Part I

Chapter 1
Introduction

Every art and every enquiry, and similarly every action and choice, is thought to aim at some good; and for this reason, the good has rightly been declared to be that at which all things aim. But a certain difference is found among ends; some are activities, others are products apart from the activities that produce them. Where there are ends apart from the actions, it is the nature of the products to be better than the activities. Now, as there are many actions, arts, and sciences, their ends are also many; the end of medical art is health, that of shipbuilding a vessel, that of strategy victory, and that of economics wealth.

Aristotle 350 BC/1998: 1094^{a}1-10

1.1 A Beginning

Above are the opening sentences of Aristotle's foundational *Nichomachean Ethics*, which more than 2350 years later than its origins remains essential reading in the serious study of ethics. These sentences stimulate two thoughts in the present context. Firstly, it is remarkable that Aristotle uses an engineering example – shipbuilding, a leading-edge technology of his time – in seeking to explain the nature of ethics. Indeed, such an example is especially relevant, for Aristotle regarded the knowledge gained by the pursuit of ethics as being "practical knowledge", a questioning and reflection that leads to action. Secondly, if we were to generalise and update the example, what would we define as being the objectives of modern engineering? In particular we might ask how we could benefit from modern ethical analysis in defining these objectives.

The present book is a response to an assessment that, despite the demonstrable achievements of modern engineering, engineers have to a significant extent forgotten that their primary objective is the promotion of human wellbeing. A crucial factor in this amnesia is that engineers have not engaged sufficiently in ethical analysis of their activities. Further, when they have so engaged they have considered only a narrow and inadequate range of the types of such analysis available. In these considerations, the activities of modern engineers do not compare favourably with the modern equivalents of the other professions referred to by Aristotle in the opening of *Nichomachean Ethics*, medicine and business. *The primary thesis of the present book is that the greatest present challenge for engineering is the development of a genuinely aspirational ethical ethos for the profession, and an outline of such an ethical ethos will be presented.*

An important purpose of the present book is to provide a conceptual framework that encourages engineers to reflect on how they can best realise the benefits of the application of their skills. In order to do so they need to allow time and effort to assess their immediate professional tasks in a broader human context. One of the reasons for the previous and current lack of such engagement is undoubtedly that the technical core of engineering is intellectually a very demanding activity. This leads to a specific danger, which has been succinctly formulated, "When one's head and hands are busy day in and day out with technology, one's heart will soon be filled with the same as well" (Schuurman 2002/2005: 17). The engineer becomes lost in "the labyrinth of technology". It is not only engineers who are dazzled by technology. The poet R.S. Thomas (1952/2000: 30) has written of a poor hill-farmer acquiring a tractor:

> Ah, you should see Cynddylan on a tractor.
> Gone the old look that yoked him to the soil;
> He's a new man now, part of the machine,
> His nerves of metal and his blood oil.

If such dazzling is experienced even on the periphery of modern society, then the engineer at the centre of technological development needs to be especially aware of his or her prioritisation and responsibilities.

1.2 What is Engineering?

The outcomes of engineering are practical. Such outcomes are most usually considered to be the design, manufacture and operation of useful devices, products and processes, often on a notably large scale. Here scale may refer to both the actual size of an individual outcome and to the number of such outcomes. To realise these outcomes, engineers use their knowledge of science and mathematics combined with imagination, reasoning, judgement and experience. The overall scope of engineering may be reviewed by considering the familiar names of some of its subdivisions, such as chemical engineering, civil engineering, electrical engineering and mechanical engineering. Each has made enormous contributions to the material wellbeing of individuals around the world. Each also creates potential risks. For example, benefits include clean water production, large-scale pharmaceutical manufacture, hygienic food processing, energy generation, buildings, transport infrastructure, mechanical devices, medical diagnostic equipment, instrumentation, computing and telecommunications. Risks include weapons manufacture and proliferation, damage to the natural environment and possible adverse effects on human health.

The practical emphasis of engineering distinguishes it from science. Simply expressed, the primary goal of science is a better understanding of the nature of the universe, and especially of the physical and biological phenomena of the world. However, the distinction is not sharp and precisely defining the boundary is not a productive activity. Thus, improved scientific understanding may often have significant practical outcomes and scientists may be motivated in their work by very practical concerns. Correspondingly, the most challenging engineering often seeks to exploit recent scientific discoveries. Indeed, engineers may themselves advance scientific understanding in their quest for better engineering solutions. It may be suggested that engineering is to science as ethics is to philosophy – both engineering and ethics are driven by a need for action, with philosophy and science being driven by a love of knowledge.

The designation "technology" hovers somewhat uncertainly between science and engineering. Technology does not primarily involve the same quest for fundamental knowledge that drives science, though technologists may contribute to basic understanding. Technology may have clear practical objectives, but these do not appear to be essential to the technologist. We are all familiar with technically ingenious gizmos that do not have a significant practical advantage or that are designed to carry out tasks that may be otherwise accomplished with greater simplicity and at lower cost. The dangers of giving priority to technical wizardry will be an important theme of the present book.

1.3 What is Ethics?

Ethics has sometimes been viewed by engineers as a somewhat arcane theoretical aspect of philosophy having little relevance to their practical activities in the world. Actually, philosophers have expressed such scepticism earlier and with intense irony, "Should philosophy, among its other conceits, imagine that someone might actually want to follow its precepts in practice, a curious comedy would emerge" (Kierkegaard 1843/1985: 124). Ethics certainly involves philosophical activities such as careful conceptual analysis and contemplation. However, ethics is in essence practical, for the way in which we choose to act and live is the primary objective of such analysis and contemplation. Ethical decisions, like engineering decisions, may have significant consequences for human wellbeing.

The outcome of ethical analysis might be considered to be a decision on how "everything ought to happen, while taking into account the conditions under which it very often does not happen" (Kant 1785/1998: 1). Ethics at its core is about how we relate to others. In such relationships, problems may arise for several reasons, including: limited resources and limited sympathy generating competition and conflict rather than mutually beneficial cooperation; limited agreement on goals and different conceptions of "good"; inadequate rationality, insufficient information and limited understanding; poor communication. Life is complex. Ethics and engineering are complex. Ethics is also, unlike much of engineering, "uncodifiable". It cannot be reduced to a "calculus of consequences" (Ratzinger 1969/2004: 27).

In conventional English usage, the designation "ethics" is to a large extent used interchangeably with the designation "morality". The origin of the word "ethics" lies in the Greek *ethikos* referring to ethos, that is, distinctive character, spirit or attitude. "Morality" comes from the Latin *moralis*, especially as used in Cicero's translations and commentaries on Aristotle, and is more concerned with which actions are right or wrong. The present book will adopt the convention of Ricoeur (1990/1992: 170) in regarding the designation "ethics" as referring to the *aims* of a life that could be described as good and regarding the designation "morality" as referring to the *rules* or *norms* that provide specific articulations of these aims. This implies the primacy of ethics over morality, but also the necessity of morality for expression of the ethical aim, though with the provision that recourse may be made to the ethical aim if the rules or norms existing at any time prove inadequate for a specific purpose. On this basis, morality expressed as norms or rules constitutes only a limited actualisation of the ethical aims. On the basis of this convention, most previous books dealing with "engineering ethics", which have largely been concerned with specific dilemmas arising in the practice of engineering, may be regarded as primarily treatments of engineering morality. Further, the several "ethical codes of conduct" for professional engineers have by their very nature tended to be prescriptions of engineering morality. A great benefit of Ricoeur's distinction in the present context is that it encourages the formulation of a genuinely *aspirational* engineering ethical ethos.

1.4 What is the Issue?

At its best, engineering changes the world for the benefit of humanity. However, there are at present significant imbalances in the application of engineering knowledge. *These imbalances reflect a tendency for engineering, as presently taught and practised, to prioritise technical ingenuity over helping people.*

In some instances, appropriate technology is available but is not being applied. A prominent example is water treatment. The provision of drinkable supplies through more effective management and treatment of freshwater resources and through desalination of sea and ground water is one of the most significant challenges that the world faces. Appropriate water management and treatment processes, both simple and advanced, are available. It might, therefore, be expected that the design, installation and operation of such processes would be accepted as being unequivocally good and would be given the highest priority. However, this is not the case. As a result, 2 billion people are affected by water shortages in over forty countries, 1.1 billion people do not have safe drinking water and 2.4 billion have no provision for sanitation. The consequences are enormous. It is estimated that 25,000 people die *every day* from water-related hunger (some specifically from thirst) and that 6,000 people, mostly children under the age of five, die *every day* from water-related diseases (UNESCO 2003).

In other instances, complex and inappropriate technology is being developed and applied. For example, the design and manufacture of military equipment and weapons would be impossible without the extensive use of advanced engineering techniques, both theoretical and practical, across the range of the engineering disciplines. World annual military expenditure is estimated to have reached $ 1339 billion in 2007 (SIPRI 2008). Almost a third of engineers in the US are reported to be employed in military related activities (Gansler 2003). Weapons are often designed to cause indiscriminate injury and death, in contravention of international conventions and treaties. For example, cluster munitions, which are a means of delivering and scattering a large number of explosive sub-munitions that due to their high failure rate remain hazardous over wide areas in large numbers for long periods of time. Children are again most often put at risk.

Peaceful and military uses of engineering compete for the same skilled personnel. Moreover, as recent conflicts show, war is often related to competition for limited resources, such as energy and water. In the case of energy, it is especially ironic that many of the engineering skills needed for civilian nuclear power are closely related to those that are essential for nuclear weapons programmes. As the availability of such skills is limited, maintenance of such weapons programmes may very significantly diminish a country's capability for achieving greater energy self-sufficiency and hence increase the likelihood of later engagement in military action to ensure future supplies.

A theme of the present book is that engineers need to think creatively and become smarter in using engineering to avert conflicts in non-military ways. An alternative to the use of engineering in preparation for military deterrence and pre-emptive war is to use the same basic skill resources in preparation for genuine peace. Thus, in the case of water, a focus on using engineering to resolve some of the intense competition for limited resources would not only resolve the immediate human need but also remove a significant likely cause of future military conflict.

In a short book it will only be possible to consider a few specific practical issues. Thus, the issues of weapons and water will be used throughout as key exemplars. Engineers among readers will be able to identify other such issues in their own fields of expertise. Present in all of the specific issues will be aspects of the changed scale of modern engineering activity compared to actions in previous human history. Whereas traditional ethics reckoned only with local and non-cumulative behaviour, our modern engineering has unprecedented and cumulative influences on multitudes of people and vast areas of the natural environment, lasting for extended periods of time. One of the first works to analyse the additional responsibility that this scale brings included the formulation of a new imperative, "Act so that the effects of your action are compatible with the permanence of genuine human life" (Jonas 1979/1984: 11), an outlook which requires special consideration in the development of an aspirational engineering ethical ethos.

1.5 Traditional Ethical Viewpoints

The documented quest for a basis for ethics has occupied philosophers for about 2500 years, though some have denied the possibility of such a basis. The present book will not provide a survey of the entire scope of this quest, for several excellent texts are available[1]. However, those approaches that have been most widely adopted by engineers will be discussed. In this discussion, the overall attitude adopted will be that which has been termed "soft objectivism", which holds that for any ethical question there *may* be a right answer (Graham 2004: 14). At the very least, this allows a rational enquiry to proceed. Of the many ways of categorising possible approaches to ethics, Graham's recent and cogent discussion identified egoism, hedonism, virtue theory, existentialism, Kantianism (duty), consequentialism, contractualism and religion. Of these approaches, those most usually adopted by engineers have been consequentialism, contractualism and duty, with some consideration of virtue based views.

There is a tendency for engineers to adopt as a default position in the consideration of ethical problems a calculation of consequences and risks, a calculus of consequences, essentially a form of utilitarianism. The most enduring account of this viewpoint is that of Mill, who defined the standard for action as "not the agent's own greatest happiness, but the greatest amount of happiness altogether" (Mill 1861/2001: 11). The approach has since been considerably sophisticated and variants of this methodology are also termed consequentialism and proportionalism. Though such assessment of consequences is important and remains a standard against which other approaches need to be compared, there appears to be insufficient awareness amongst engineers of its fundamental difficulties as identified by philosophers of very diverse viewpoints. Some of these difficulties and their possible resolution will be analysed.

Contractualism regards the motivation for acting ethically as being social agreement backed by a regulative framework. A celebrated early version of such a contract is developed by Socrates in Plato's dialogue *Crito*, where it is said to be established between the laws of the city and each citizen (Plato c.395 BC/1997). The approach has been refined in many ways since, with the most significant modern work being that of Rawls, who claimed to provide, "an alternative systematic account of justice that is superior ... to the dominant utilitarianism of the tradition" (Rawls 1971/1999: xviii), and where the emphasis is on the development of a contract between individuals. Professional ethical codes of conduct are by their very origin and nature likely to have a strong contractual aspect. The present book will analyse some of the benefits of a contractual approach to engineering ethics and also some of the limitations, many of which are related to the international and multicultural nature of the profession. It will be proposed that

[1] For example, Graham (2004), Vardy and Grosch (1999).

a crucial limitation in the present context is that contractualism provides more of a basis for securing the present arrangements, even of justifying ethical mediocrity, rather than providing an opportunity for promoting an ethos of high ethical aspirations. In addition, international agreements are frequently not adhered to.

An important aspect of the aspirational engineering ethos to be outlined in the present book is an emphasis on our responsibility for others. A foundational formulation of the priority of respect for others in Western philosophical ethics has been given by Kant. Kant was especially concerned with actions undertaken from duty, that is because they are the right things to do rather than because they are an immediate inclination or because they further some end. He is famous for the statement of the "categorical imperative", "act only in accordance with that maxim through which you can at the same time will that it become a universal law" (Kant 1785/1998: 31). However, he emphasised that this fundamental law required a respect for persons and hence reformulated it as, "So act that you use humanity, whether in your own person or in the person of any other, always at the same time as an end, never merely as a means" (Kant 1785/1998: 38). The respect for persons inherent in a duty approach has much to offer engineering ethics, but the often dense associated argumentation in its modern expression is a barrier to its more widespread adoption by engineers as an underlying basis for their actions.

As previously mentioned, Aristotle considered ethics to be a practical pursuit and his work remains very relevant to the practice of engineering. He was greatly concerned to identify the factors that can promote human flourishing. His approach to ethics is termed "virtue theory", as the qualities of mind and character that promote such flourishing are central to his thinking. For Aristotle, friendship was the greatest aim of the ethical life. He noted that "friendship asks a man to do what he can, not what is proportional to the merits of the case" (Aristotle 350 BC/1998: 1163^{b}15), which certainly provides a succinct possible formulation of an aspirational ethos appropriate for engineers. In the present book, an important modern analysis and development of virtue theory by MacIntyre will provide essential features of the aspirational engineering ethos to be developed.

1.6 Other Professions: Medicine and Business

In the opening section of the *Nichomachean Ethics*, Aristotle referred not only to engineering, shipbuilding, but also to medicine and business. A consideration of ethical practices in the modern professions of medicine and business can be very beneficial in the preparation of an aspirational engineering ethos.

Modern medicine uses very advanced science and engineering. Indeed, without engineering medicine would be a very restricted activity – there would be no advanced building design, sophisticated diagnostic equipment or pharmaceuticals.

Yet, in contrast to engineering, medicine is seen as the paradigm of a caring profession. This difference appears to be the consequence of two main factors:

(i) Medicine is seen as relating more directly to people. In some ways this is a curious perception, for most people generally only come into contact with a doctor if they are ill whereas virtually every aspect of our lives is affected by engineered artefacts or processes. However, the engineer often does not come into direct contact with the people affected by his or her work and, due to the complex nature of engineering projects, there may be a significant time lapse between his or her initial work and the final outcome of his or her activities. It will be proposed that redressing this lack of physical and temporal proximity is an important aspect of promoting an ethical approach to engineering.
(ii) Medicine has a vigorous and public ethical debate about its activities whereas this has been lacking for engineering. This lack of debate, among both engineers and philosophers, is surprising considering the great impact that the practice of engineering has on our lives. The present book aims to stimulate such ethical debate among both engineers and philosophers, and also more widely in society.

Two features of modern medical ethics hold special lessons for the development of an aspirational engineering ethical ethos. Firstly, there is a great emphasis on the *individual*, the patient affected by the doctor's actions. Secondly, the doctor is emphatically *personally* accountable for decisions.

Business ethics is prominent both within business and in public awareness. Engineering and business have a symbiotic relationship, for engineers need to work with those who have business skills in order to exploit their technical skills, and business needs the skills of engineers for the design and manufacture of saleable products. One of the important ways in which business ethics can help the formulation of an aspirational engineering ethical ethos is though its analysis of the situation of individuals employed in large enterprises. Two related aspects are important. First, there is a lack of freedom for many staff in large organisations. They are employed for the way in which their know-how promotes the achievement of certain specified ends so that their discretion, where it exists, often lies only in the choice of means. This can lead to a suppression of ethical responsibility, a bracketing of personal ethical values. Second, there is a tendency for individuals to be less ethically alert in their role as employees than they are as private persons. Social science studies suggest that employees often function at a level of minimally acceptable ethical behaviour rather than aspiring to the highest achievable levels. In response to this first aspect it will be appropriate to consider how possible changes to working practices and career structures can give engineering employees a greater awareness of ends. Redressing the second aspect will require a consideration of how an engineer's professional and private concerns can be integrated in a more coherent way.

1.7 Outline of an Aspirational Engineering Ethics

An aspirational engineering ethical ethos needs to overcome the limitations of the traditional ethical views and to learn from the more advanced analysis of ethics in other professions. The present book builds on two philosophical sources. Firstly, on writings that lie somewhat outside what is conventionally regarded as the mainstream of ethics, but which contain profound ethical insights that can provide an important balance to prevailing views. Secondly, on recent writings that build on the philosophical mainstream in especially imaginative and practical ways.

Foremost among writings in the first category is Buber's *I and Thou* (1923/2004), which is concerned with the way in which relation arises. Buber introduced a seminal vocabulary for describing our existence by using the "primary words" *I-Thou* and *I-It*. The world of *I-It* is the world of *experiencing* and *using* where a man or woman "works, negotiates, bears influence, undertakes, concurs, organises, conducts business, officiates ..." (Buber 1923/2004: 39). Engineers will recognise the *I-It* world as a description of the conditions under which they usually carry out their professional activities. In contrast, the world of *I-Thou* is the world of *meeting* nature and people, with the possibility of a relationship engendering care. This is an aspect lacking in most analyses of modern engineering. Indeed, Buber warned that vigilance is necessary in highly technological societies, for he had observed that the development of experiencing and using leads to a decrease in the ability to enter into relation. For the formulation of an aspirational ethical ethos for engineering, Buber's insight has the unique advantages of encompassing both person/person and person/natural world (environmental) relations and of recognising the importance of technical knowledge. It also vitally balances the priority presently given to rule and outcome approaches in engineering ethics. However, the nature of engineering requires an extension of *I-Thou* interactions in terms of *I-You* interactions, based on care but lacking personal proximity.

Levinas provided an even stronger statement of the priority of the demands that others make on us, which he designated by the strikingly visual notion of *the face*. He describes an ethical act as, "a response to the being who in a face speaks to the subject and tolerates only a personal response" (Levinas 1961/1969: 219). In simple terms, and changing the metaphor, we need to hear the voice of others saying, "It's me here, please help me!". The directness of the approaches of both Buber and Levinas make them very appropriate starting points for the development of an aspirational engineering ethos. It is worth noting that Korsgaard (1996) has reached a similar conclusion about the ethical priority of others through a sophisticated Kantian-duty based analysis of a type that only the most ethically committed of engineers are likely to follow.

Of works in the second category, MacIntyre's *After Virtue* (1981/1985) provides further key concepts for the development of an aspirational engineering ethical ethos. Like Buber and Levinas, he argues that traditional approaches have placed too little emphasis on persons, and additionally that such approaches also placed too little emphasis on the contexts of their lives. In his view, the ability to give ethical priority to people is facilitated by cultivation of personal virtues, which may depend on context. This approach is promoted by consideration of the appropriate virtues in terms of a *practice* supported by *institutions*. Though he does not mention engineering, MacIntyre's virtue based description of practices and institutions can provide the starting point for a coherent description of engineering that may be readily recognised by professional engineers. The virtues (or principles) particularly appropriate to engineering may be categorised, for example, as: accuracy and rigour; honesty and integrity; respect for life, law and the public good; responsible leadership, listening and informing (RAE 2007). MacIntyre's terminology also provides a very appropriate description of the outcomes of engineering activity. These are *internal goods*, including standards of technical excellence and the satisfaction arising from personal accomplishment, and *external goods*, including engineered artefacts and wealth. The description of engineered artefacts as *goods* allows such physical technological accomplishments to be distinguished from the *end* or *goal* of engineering activity, which may be described as the promotion of human flourishing through contribution to material wellbeing. *Such an analysis leads to the important conclusion that the present imbalanced prioritisation in engineering of technical ingenuity over helping people may be considered as arising from mistaking the external goods of the practice for the real end of the practice.*

The development of an aspirational ethical ethos for engineering requires an antidote to the tendency of individuals to adopt lower ethical aspirations in professional contexts than in private contexts. It will be proposed that this may best be achieved through encouraging the consideration of life as a *narrative unity*, including the aim of coherence in ethical standards in all activities. Such an approach further promotes ethical aspirations through generating an awareness of the desire of others to achieve a narrative unity in their lives. This leads to the need to consider how a balance may be achieved in *I*, *I-Thou* and *I-You* prioritisation, or alternatively expressed, in self care, proximate care and care beyond proximity. Achieving a balance when prioritising such commitments is especially challenging for engineers as their professional activities may have substantial influence far beyond proximity in both place and time. Two factors are especially significant in such a balance. Firstly, an acknowledgement that there is a limit to the extent that any individual can prioritise ethical behaviour. Very few people are saints, and indeed sainthood may indicate a lack of a full range of human values and activities. Secondly, the need for compassion and generosity in professional as well as in private life. Such a balance requires sensitivity, including sensitivity to consequences. However, a *calculation* is never possible, a *decision* has to be made.

1.8 Practical Outcomes

The outcomes of both engineering and ethics are practical. Hence, any analysis of engineering ethics should be doubly expected to have practical outcomes. Therefore, the final task of the present book will be to consider some of the practical outcomes of the analysis presented. These outcomes relate to ways in which an aspirational ethical ethos for the profession may be promoted, rather than to the solution of specific moral quandaries. Important outcomes to be discussed include:

In education Recruitment of students and the overall tone of engineering courses should give greater emphasis to the goal of benefits in terms of the quality of life. Personal responsibility for the results of professional activities should be emphasised.

Engineering institutions The professional codes of national engineering institutions should progressively incorporate increased degrees of compassion and generosity. The development of an effective international engineering initiative to promote aspirational practice should be supported.

Industry and work practices The more widespread adoption of aspirational codes of conduct in industry should be promoted. Career development plans that bring employees into closer proximity with end-users for at least part of their working life should be encouraged.

Positioning engineering in the public and intellectual mainstreams The public should be engaged in positive debate about engineering priorities, as already occurs for medicine and business.

An aspirational role for engineering in international initiatives Alignment of engineering aspirations with international initiatives such as the United Nations *Declaration and Programme of Action on a Culture of Peace* may be an especially effective way forward. There is also a need to prioritise a culture of peace *within* engineering.

The personal ethical responsibility of every engineer All engineers need to take a more active role in considering the ethical implications of their work. Our aspiration should be summarised: *"Here I am, how can I help you?"*.

References

Aristotle (1998) Nichomachean ethics. In: The complete works of Aristotle, vol 2, J. Barnes (ed), Princeton University Press, Princeton. Original attribution 350 BC

Buber M (2004) I and Thou, Continuum, London. Originally published as Ich und Du, Shocken Verlag, Berlin, 1923

Gansler JS (2003) Integrating civilian and military industry, Issues in Science and Technology. http://www.issues.org/19.4/updated/gansler.html
Graham G (2004) Eight theories of ethics, Routledge, London
Jonas H (1984) The imperative of responsibility, Chicago University Press, Chicago. Originally published as Das Prinzip Verantwortung (1979) Insel Verlag, Frankfurt am Main
Kant I (1998) Groundwork of the metaphysics of morals, Cambridge University Press, Cambridge. Originally published as Grundlegung zur Metaphysik der Sitten, 1785
Kierkegaard S (1985) Fear and trembling, Penguin Books, London. Originally published as Frygt og bæven, 1843
Korsgaard CM (1996) The sources of normativity, Cambridge University Press, Cambridge
Levinas E (1969) Totality and infinity, Dudesque University Press, Pittsburgh. Originally published as Totalité et infini, Martinus Nijhoff, The Hague, 1961
MacIntyre A (1985) After virtue, 2nd edition, Duckworth, London. First published 1981
Mill J S (2001) Utilitarianism, Hackett Publishing Company, Indianapolis. First published 1861
Plato (1997) Crito, in "Plato: Complete works", Cooper JM (ed) Hackett Publishing Company, Indianapolis. Original attribution c. 395 BC
Ratzinger J (2004) Introduction to Christianity, new edition, Ignatius Press, San Francisco. Originally published as Einführung in das Christentum, Kösel-Verlag GmbH, Munich, 1969
Rawls J (1999) A theory of justice, revised edition, Oxford University Press, Oxford. First published 1971
Ricoeur P (1992) Oneself as another, University of Chicago Press, Chicago. Originally published as Soi-même comme un autre, Editions du Seuil, 1990
Royal Academy of Engineering (2007) Statement of ethical principles, RAE, London
Schuurman E (2005) The technical world picture and an ethics of responsibility, Dordt College Press, Sioux Centre. Originally published as Bevrijding van het technische wereldbeeld. uitdaging tot een andere ethiek, Technische Universiteit, Delft, 2002
Stockholm International Peace Research Institute (2008) SIPRI Yearbook 2008: Armaments, disarmament and international security, Oxford University Press, Oxford
Thomas RS (2000) Collected poems 1945–1990, new edition, Phoenix, London. Poem first published 1952
United Nations Educational, Science and Cultural Organisation (2003) World water development report: water for people, water for life, UNESCO, Paris
Vandenburg WH (2000) The labyrinth of technology, University of Toronto Press, Toronto
Vardy P, Grosch P (1999) The puzzle of ethics, revised edition, Fount Paperbacks, London

Chapter 2
What is the Issue?

2.1 Introduction

Engineering has made an enormous contribution to providing the material well-being that promotes human flourishing. The everyday benefits of engineering include the provision of energy, clean water, sanitation, hygienic food production, pharmaceutical manufacture, buildings, transport, communications and computers. Many of these benefits have become so closely integrated with our everyday life that we are often unaware of our dependence on them until a failure occurs. However, it takes only a relatively brief interruption in the provision of these benefits for us to realise again how important they are to the maintenance of our quality of life.

The benefits of engineering have been documented in many articles, books, radio programmes and television documentaries. For example, in 2005 the prestigious BBC Reith Lectures were given by the then President of the UK Royal Academy of Engineering, with the title, *The Triumph of Technology* (Broers 2005). He outlined how successful application of engineering has, from the beginnings of civilisation, greatly improved standards of living. He gave examples ranging from Neolithic mining to recover flints for the production of sharp cutting edges – literally "cutting edge technology" – to the latest advances in nanotechnology. His examples amply demonstrated that engineering allows more and more people to enjoy a possibility of leading fulfilled lives in ways that were once restricted only to the wealthy.

At its best, engineering has changed the world for the benefit of humanity in remarkable ways. Even so, there are darker sides to engineering, as there are with most human activities. An important purpose of the present book is to provide a conceptual framework that encourages engineers to reflect on how they can optimise the benefits of the application of their skills. As engineering is practical, it can be helpful at the outset to consider some cases where application of engineering skills is either directly deleterious or otherwise clearly less than optimal so as to better appreciate why a new conceptual framework in needed. As an example of the first type, the development of indiscriminate weapons systems will be considered. As an example of the second type, the ineffectual application of processes for drinking water provision and sanitation will be described. It will be further shown that such examples are closely linked. Continued weapons development can produce a transient peace, but it is essentially an example of trying to solve a problem by the same type of thinking that created it. We should be able to do better than this. It would be smarter to use engineering for the non-military prevention and resolution of conflicts. We could, for example, focus our skills more directly on resolving some of the intense competition for limited resources, such as energy and water, that is likely to result in future conflicts. Even more positively, we could use engineering for the promotion of a culture of peace. Finally, the present chapter will consider some of the more general aspects of the changed scale of modern engineering activity compared to actions in previous human history.

2.2 Military Technology

World annual military expenditure is estimated to have reached $1339 billion in 2007 (SIPRI 2008). The design and manufacture of military equipment and weapons would be impossible without the extensive use of advanced engineering, both theoretical and practical, across the range of the engineering disciplines. The extent of engineering skills committed to the development, manufacture and use of such technologies is enormous. For example, almost a third of engineers in the US are reported to be employed in military related activities (Gansler 2003). Companies deriving much of their sales income from military technology are also major employers of engineers in the UK.

In a world of limited resources, limited sympathy and limited rationality, competition leading to conflict can arise rather than a more desirable state of mutually beneficial cooperation. In response to this state of affairs, theologians, philosophers and politicians have sought to limit the scope of such conflict through written analysis and international agreements. The present section will consider some of these approaches with reference to selected engineered weapons systems.

2.2.1 "Just" War

Recognising that the "melancholy state of humanity" makes the possibility of war a constant reality, theologians and philosophers have from early times attempted to formulate conditions which might lead to the avoidance or limitation of armed conflict. The many formulations have subtle differences, but a representative account itemises five requirements dealing with the decision to *commence* war (Jones 1998):

1. There must be a just cause (such as to repel an aggressor).
2. There must be a just intent (such as to restore peace and justice).
3. War must be a last resort, every possibility of peaceful settlement having been exhausted.
4. The declaration of war must be by a legitimate authority.
5. There must be a good prospect of success.

and a further two requirements for the *conduct* of war:

6. The innocent must not be directly attacked, but only the armed forces of the enemy.
7. The means used must be in proportion to the end in view.

These requirements provide a useful basis for analysis and discussion, but they have also been subject to strong criticism. For example, the very concept of a "just" war may be questioned, for armed conflict inevitably results in the death and injury of the innocent. Indeed, it is estimated that ninety percent of those killed, wounded, abused or displaced in violent conflict are (civilian) women and children (UNF 2008). The requirements represent a rather idealised state very different from the irrational political disputes that precede the start of wars and the chaotic conduct of wars. It has also been suggested that the very existence of a set of requirements of this sort makes war more likely, as they can be used to justify war.

The two requirements for the *conduct* of war, the use of discriminate and proportional means, are those which are most immediately relevant to the use of engineering for military purposes. However, in a broader engineering context, the third requirement for *commencement* of war, that it must be a last resort with all non-military options having been exhausted, is even more important. The application of these requirements to specific weapons systems will be considered later in the present chapter.

2.2.2 International Conventions and Treaties[1]

Recognition of the undesirability of war has led to the formulation of a number of significant international conventions and treaties, which are complemented by extensive diplomatic discussions. A systematic account of these activities is beyond the scope of the present book. However, two which are especially relevant to the

[1] The full texts of relevant international conventions and treaties are available on the websites of international organisations such as the International Committee of the Red Cross, the International Atomic Energy Agency and the United Nations.

engineered weapons systems to be considered later require attention: the *Protocol Additional to the Geneva Conventions of 12 August 1949, and relating to the Protection of Victims of International Armed Conflicts* (Protocol 1), *8 June 1977* and the *Treaty on the Non-proliferation of Nuclear Weapons* (NPT) which was first signed on 1 July 1968 and entered into force on 5 March 1970.

An important aspect of Protocol 1 is the protection of the civilian population in the event of hostilities. The basic rule is set out in Article 48:

> In order to ensure respect for and protection of the civilian population and civilian objects, the Parties to the conflict shall at all times distinguish between the civilian population and combatants and between civilian objects and military objectives and accordingly shall direct their operations only against military objectives.

In concurrence with the requirements for the conduct of a "just" war, Article 51 (4) specifies:

> Indiscriminate attacks are prohibited. Indiscriminate attacks are:
> (a) those which are not directed at a specific military objective;
> (b) those which employ a method or means of combat which cannot be directed at a specific military objective; or
> (c) those which employ a method or means of combat the effects of which cannot be limited as required by this Protocol.

There is a further requirement for proportionality, prohibiting:

> An attack which may be expected to cause incidental loss of civilian life, injury to civilians, damage to civilian objects, or a combination thereof, which would be excessive in relation to the concrete and direct military advantage anticipated. (Article 51 (5b))

The Protocol also specifies that civilian objects shall not be the subject of attack and reprisals, specifying in particular the protection of objects indispensable to the survival of the general population:

> It is prohibited to attack, destroy, remove or render useless objects indispensable to the survival of the civilian population, such as food-stuffs, agricultural areas for the production of food-stuffs, crops, livestock, drinking water installations and supplies and irrigation works, for the specific purpose of denying them for their sustenance value to the civilian population or to the adverse Party, whatever the motive, whether in order to starve out civilians, to cause them to move away, or for any other motive. (Article 54 (2))

Precautionary measures are also required:

> In the conduct of military operations, constant care shall be taken to spare the civilian population, civilians and civilian objects. (Article 57 (1))

Protocol 1 has been widely accepted internationally, though it has not been ratified by several countries currently engaged in military conflicts.

The NPT is concerned in particular with nuclear weapons, "Considering the devastation that would be visited upon all mankind by a nuclear war and the consequent need to make every effort to avert the danger of such a war and to take measures to safeguard the security of peoples". The NPT consists of eleven articles, four of which are especially relevant in the present engineering context. Articles I and II deal with non-proliferation:

> Each nuclear-weapon State Party to the Treaty undertakes not to transfer to any recipient whatsoever nuclear weapons or other nuclear explosive devices or control over such weapons or explosive devices directly, or indirectly; and not in any way to assist, encourage, or induce any non-nuclear weapon State to manufacture or otherwise acquire nuclear weapons or other nuclear explosive devices, or control over such weapons or explosive devices. (Article I)

> Each non-nuclear-weapon State Party to the Treaty undertakes not to receive the transfer from any transferor whatsoever of nuclear weapons or other nuclear explosive devices or of control over such weapons or explosive devices directly, or indirectly; not to manufacture or otherwise acquire nuclear weapons or other nuclear explosive devices; and not to seek or receive any assistance in the manufacture of nuclear weapons or other nuclear explosive devices. (Article II)

Article IV is concerned with the availability of nuclear technology for peaceful purposes:

1. Nothing in this Treaty shall be interpreted as affecting the inalienable right of all the Parties to the Treaty to develop research, production and use of nuclear energy for peaceful purposes without discrimination and in conformity with articles I and II of this Treaty.
2. All the Parties to the Treaty undertake to facilitate, and have the right to participate in, the fullest possible exchange of equipment, materials and scientific and technological information for the peaceful uses of nuclear energy. Parties to the Treaty in a position to do so shall also cooperate in contributing alone or together with other States or international organizations to the further development of the applications of nuclear energy for peaceful purposes, especially in the territories of non-nuclear-weapon States Party to the Treaty, with due consideration for the needs of the developing areas of the world.

Article VI is very sweeping in scope, relating to nuclear disarmament and complete disarmament:

> Each of the Parties to the Treaty undertakes to pursue negotiations in good faith on effective measures relating to cessation of the nuclear arms race at an early date and to nuclear disarmament, and on a Treaty on general and complete disarmament under strict and effective international control.

As with Protocol 1, there is wide international assent to NPT, though also significant controversy about its application. Article VI has so far had little effect, and it notably omits any timescale, though the Review Conference of the NPT in 2000 agreed on a series of thirteen practical steps for the systematic and progressive implementation of this article.

In the context of Protocol 1 and the NPT, the following section will consider two weapons systems based on advanced engineering, cluster munitions and nuclear weapons. A later section will consider the broader context of the most effective use of engineering for the promotion of peace and prosperity including the responsibility of engineers to make a contribution complementing and enhancing the political and diplomatic endeavours that have led to the wide range of international conventions and treaties.

2.2.3 Cluster Munitions and Nuclear Weapons

A cluster munition is a means of delivering and scattering a large number of explosive sub-munitions. A cluster bomb might typically deliver about 150 bomblets, each containing 100 g of high explosives wrapped in a coil of 2,000 pieces of shrapnel. More recently there has been a move to deliver cluster munitions using missile and artillery projectiles. A single cluster munition may scatter sub-munitions over an area of one square kilometre. The number of sub-munitions delivered by such means may be enormous: during the 1991 US operation "Desert Storm" it is estimated that 11,000,000 were fired from rockets, 220,248 in the *first five minutes* (McGrath 2004).

Cluster sub-munitions cause extensive damage. They cause immediate indiscriminate injury or death to civilians. They use shrapnel made of materials that are difficult to detect during medical treatment. Their "failure" rate, which may be in the range 5–30%, ensures that they remain hazardous in large numbers for long periods of time. Due to their small size, such "failed" sub-munitions are especially attractive, and hence hazardous, to children. Their dispersal and difficulty of detection makes them especially lethal in built-up areas and can render large areas of agricultural land unusable. For example, the United Nations reported that, following the month-long conflict in 2006, there were 100,000 unexploded bomblets in south Lebanon spread over 359 separate locations, which had been fired by Israeli forces – 90% of strikes occurring in the last 72 hours of conflict when a resolution was known to be imminent (UN News Centre 2006). Sub-munitions dropped in Vietnam and Laos by US forces in the period 1961–1975 are reported to still cause hundreds of injuries and deaths to the civilian population each year.

The design, manufacture and delivery of cluster munitions clearly require the application of advanced engineering across a wide range of the engineering disciplines. Their high "failure" rate is especially noteworthy. Such a failure rate would be unacceptable in any other engineered product, so it is very likely that they are designed to "fail". They do not meet the requirements for the conduct of a "just" war and contravene Protocol 1: they are indiscriminate in their immediate effect; perversely, they are later most likely to cause injury to civilians rather than military personnel; they remain dangerous for disproportionate lengths of time; when dispersed over agricultural land they deprive targeted populations of means of sustenance. Cluster munitions have been used in at least twenty-three countries and stockpiles totalling billions of sub-munitions are held by an estimated seventy states. *Complex but inappropriate technology is being developed and applied – illegally – on a huge scale, and a large part of the responsibility for this development and application rests with engineers*[2].

The NPT recognises the unique scale of the devastation caused by nuclear weapons even in comparison to the most deadly of conventional weapons. As their use has shown, nuclear weapons cause massive and indiscriminate destruction with consequences affecting civilian populations for decades. The commitment of engineering skills and resources in the design, manufacture, maintenance and deployment of these weapons is proportionately massive. The US is reported to have a force of 14 submarines each carrying 24 missiles, 500 land-based inter-continental ballistic missiles, air-delivered cruise missiles and free-fall bombs delivered by a range of aircraft. Russia has more than 250 submarine-launched cruise missiles, 520 inter-continental ballistic missiles, about 700 air-launched cruise missiles and a large number of non-strategic nuclear weapons. France has total warhead numbers around 350 and Britain around 200. China has around 20 inter-continental ballistic missiles and a larger number of nuclear-armed intermediate and medium range ballistic missiles. India, Pakistan, Israel and North Korea also have nuclear weapons (Cabinet Office 2006). All of these countries are developing new nuclear weapons systems, requiring a considerable expenditure in research and development. *Nuclear weapons are another example of the development and application of complex but inappropriate technology on a huge scale, and again a large part of the responsibility rests with engineers.*

2.3 Water

There is strong historical evidence that one of the most effective ways in which engineering can contribute to human wellbeing is through the provision of safe drinking water and adequate sanitation. For example, in the 1840s infant mortality in Great Britain was about 160 deaths for every 1000 live births, roughly the same

[2] (i) Zak Johnson, who has over twenty years experience of military and humanitarian explosive ordinance disposal, including work in Afghanistan and Laos, raises the following issue, "I have often wondered why intelligent engineers use their skills to design and manufacture more effective ways to kill and injure people. It must be a strange way to make a living and what do you tell your family?" (private written communication).
(ii) An initiative to limit the development, production, acquisition and use of certain types of cluster munitions is in progress at the time of writing: a *Convention on Cluster Munitions* was endorsed in May 2008 and is due to be signed in December 2008. This is a very welcome development, but the Convention is opposed by several states that are key producers and users of such weapons. Cluster munitions are reported by the UN and humanitarian organisations to have been used in the conflict in South Ossetia in August 2008.

as in the poorest countries today[3] (UNESCO 2006. Other data in this paragraph are from the same source). Children died especially from diarrhoea and dysentery. Realisation that lack of access to affordable safe drinking water and adequate sanitation were important causes of mortality led to a number of reforms in the following decades. Initial emphasis on the public provision of water did not lead to much reduction in mortality as sewage was not adequately separated from drinking water sources. However, when public investment in sanitation was also increased, peaking around 1900, there was a rapid reduction. In just over a decade from 1900 the infant mortality rate fell from about 160 to 100, one of the steepest declines in history. Similar health benefits were achieved in the US at about the same time. For example, around 1900, infectious diseases accounted for 44% of mortality in US cities, with waterborne diseases such as typhoid, cholera and diarrhoea responsible for a quarter of such deaths. Municipalisation leading to more careful sourcing of water and disposal of waste were important factors in improving the situation. However, a key role was played by the introduction of water filtration to remove bacteria. For example, typhoid mortality in Cincinnati fell from 60 per 100,000 people to 5 per 100,000 people in about two years following the introduction of filtration, with chlorination later giving further reduction.

A range of effective processes, both simple and advanced, is now available for the production of safe drinking water and for sanitation. Given the proven effectiveness of such technologies for improving health, it might be expected that the design, installation and operation of such processes would be given the highest priority. However, this is not the case. As a result, 2 billion people are affected by water shortages in over forty countries, 1.1 billion people do not have safe drinking water and 2.4 billion have no provision for sanitation (UNESCO 2003. Other data in this paragraph are from the same source). The consequences are enormous. It is estimated that 25,000 people die *every day* from water-related hunger (some specifically from thirst) and that 6,000 people, mostly children under the age of five, die *every day* from water-related diseases. For example, diarrhoeal diseases kill about 2 million people each year, including an estimated 1.4 million children under five.

Recognizing the need to speed up poverty alleviation and socio-economic development, the 2000 UN General Assembly Millennium Meeting established eight Millennium Development Goals (MDG) with targets to be achieved by 2015: eradicate extreme poverty and hunger; achieve universal primary education; promote gender equality and empower women; reduce child mortality; improve maternal health; combat HIV, AIDS, malaria and other diseases; ensure environmental sustainability; develop a global partnership for development (UN 2000). Water engineering can play an important role in achieving all of these goals, even those that appear in the first instance to be societal. For example, children who do

[3] Infant mortality for the EU is now 5.0 and for the US it is 6.5, for every 1000 live births. Infant mortality is the death of infants in the first year of life.

not suffer from water borne diseases can benefit fully from available education, and better access to water reduces the work burden on women, who are primarily responsible for seeking water in poor societies.

The costs and benefits of water and sanitation improvements at a global level have been evaluated (Hutton and Haller 2004). Five levels of intervention were considered:

Intervention 1 Halving the proportion of people who do not have access to improved water resources by 2015, a MDG target, with priority given to those already with improved sanitation. ("Improved water" includes basic technology such as stand post, borehole, protected spring or well, or collected rainwater).

Intervention 2 Halving the proportion of people who do not have access to improved water resources *and* improved sanitation facilities by 2015, a MDG target. ("Improved sanitation" includes septic tank, pour-flush, simple pit latrine, small bore sewer or ventilated improved pit latrine.)

Intervention 3 Access for all to improved water and sanitation.

Intervention 4 A minimum of water disinfected at the point of use for all (chlorine), on top of improved water and sanitation services.

Intervention 5 Access for all to a regulated piped water supply and sewage connection into their houses.

All interventions were compared to the situation in 2000 and costed to include full investment and annual running costs. The total annual costs were estimated to be: *Intervention 1* – $ 1.8 billion; *Intervention 2* – $ 11.3 billion; *Intervention 3* – $ 22.6 billion; *Intervention 4* – $ 24.6 billion; *Intervention 5* – $ 136.5 billion. There are, of course, many uncertainties in such calculations and detailed country case studies are required. Nevertheless, the simplest interventions are very low cost compared to an annual global military expenditure of $ 1339 billion, and in such comparison even the cost of the most expensive intervention is modest.

It is easy to lose an awareness of individuals in global calculations of this type, so it is helpful to consider the annual cost per person receiving interventions: *Intervention 1* – $ 2.8; *Intervention 2* – $ 5.6; *Intervention 3* – $ 8.0; *Intervention 4* – $ 4.3; *Intervention 5* – $ 22.9. These are very modest sums in comparison with the benefits to the individuals. Benefits include time savings due to better access to water and sanitation facilities, gain in productive time due to less time spent ill, health sector and patient costs saved due to less treatment of diseases and the value of prevented deaths. Even under pessimistic scenarios the potential economic benefits generally outweighed the costs.

Implementation of these interventions would be subject to political and institutional constraints in addition to the economic requirements. *However, from an engineering perspective, appropriate technology is available but is not being applied as extensively or as rapidly as it could be. This should be of great concern to engineers – they have the skills to alleviate an enormous burden of human suffering at modest cost, but are constrained from taking effective action.*

2.4 Engineering for the Promotion of Peace and Prosperity

In 2006 the UK government decided to develop a new nuclear weapons system at an estimated initial cost of £ 20 billion. In the prior debate, the *Independent* newspaper (31 October 2005) asked two pertinent questions, "Is this a sensible way to spend £ 20 billion? What else could you buy?". Among the suggestions related to engineering were: 4 Channel tunnels; 60 medium sized hospitals; 800 secondary schools; 400 trains for the London Underground; 715 miles of motorway. The scale of these alternatives for what is a small proportion of UK military spending, and a minor amount of world military spending, again indicates how the use of engineering has been diverted from those instances where it can most benefit human wellbeing.

It may be accepted that, in a world comprising nation states, a key responsibility of any government is to ensure the security of its citizens. Hence, in the present context, another pertinent question may be asked, "Is a new nuclear weapons system the most effective means of using engineering to promote peace and prosperity?". The UK government's case for a new nuclear weapons system emphasised the need to insure against an uncertain future. Four main uncertainties were identified: (i) the continued and unpredictable proliferation of nuclear weapons, (ii) weak and failing states offering safe havens for international terrorists and creating wider instability, (iii) increasing pressure on key resources such as energy and water which may increase interstate tension, (iv) the rapid and uncontrollable development of military–relevant technology by the civil sector making potential adversaries increasingly capable (Cabinet Office 2006: 18). All of these uncertainties have an engineering component. However, to seek to respond to them by further military expenditure is a clear example of the unproductive tendency of trying to solve problems by the same type of thinking that created them. We should be able to find better responses.

The uncertainty most amenable to a peaceful engineering solution is the pressure on energy and water. Major current conflicts and international disputes are related to energy resources, especially oil and gas. A country or group of cooperating countries that are self-sufficient in energy would be more likely to be free from such conflicts and disputes. In this case, it is especially ironic that many of the engineering skills needed for civilian nuclear power are closely related to those needed for nuclear weapons programmes. As the availability of such skills is limited, a new nuclear weapons programme could very well significantly diminish the

UK's capability for achieving greater energy self-sufficiency and hence increase the likelihood of later engagement in military action to ensure future supplies. Tensions related to water are already apparent between a number of nation states including: Israel, Jordan and Palestine; Syria and Turkey; China and India; Egypt and Ethiopia; Angola and Namibia. Engineering solutions to water management problems are readily available at a very modest cost compared to military solutions. Inter-state tension may be peacefully avoided if governments and international organisations prioritise water and sanitation provision and distribution. Regrettably, the present prioritisation is inadequate. Additionally, water management technologies provide real commercial opportunities allowing enhancement of economic prosperity as well as promoting peace.

The other main security uncertainties also have engineering aspects. Proliferation would be impossible without the use of engineering skills. Unstable states and international terrorists pose the greatest danger when they have access to advanced weapons technology, and access to such technology becomes easier as military expenditure generally increases and weapons become more readily available. The adaptation of civilian technology for military purposes also requires the cooperation of engineers.

Overall, the actions of engineers can have a major influence on the conditions for conflict or of international peace and prosperity. It is regrettable that they are involved in the production and use of weapons that are contrary to international law. It is regrettable that many do not develop their professional activities in response to aspects of international agreements such as Article VI of the NPT, regarding the cessation of the nuclear arms race and complete disarmament, or even of Article IV, regarding the availability of nuclear technology for peaceful purposes. Too little attention is given to the use of engineering for the non-military prevention and resolution of conflict. One of the requirements for a "just" war, that war must be a last resort, has certainly not been met until such non-military means have been exhaustively used. The great magnitude of world-wide military budgets itself indicates that war is not being considered as a last resort, but rather as a normal part of business and politics. The actions of engineers also have a significant influence on the nature of civil society. Countries with high military spending tend to have high levels of internal violence and social exclusion, as demonstrated by the high level of violent crime and the surprising extent of poverty in the US.

An important purpose of the second part of the present book is to provide a conceptual framework that encourages engineers to reflect on how they can optimise the benefits of the application of their skills and hence realise their primary objective of promoting human wellbeing. Some specific examples of the scope of engineering and of the issues involved have already been presented. However, before proceeding further it is worth considering in a more general way the changed scale of modern engineering activity compared to actions in previous human history.

2.5 The Scale of Modern Engineering Activity

Although some aspects of the immediately preceding discussion move toward the political, it will be a contention of this book that engineers themselves should take responsibility for the nature and outcomes of their professional activities. Karl Marx (1845) famously wrote, “The philosophers have only interpreted the world, in various ways; the point is to change it”. Engineers really do change the world, and on a scale unprecedented in human history. In extreme cases, nuclear weapons being one example[4] and global warming possibly another, the very continuation of human life is threatened by engineering activities. It was this that led Jonas (1979/1984: 11) to formulate a new general imperative, “Act so that the effects of your action are compatible with the permanence of genuine human life”.

The extent of the influence of modern engineering is greatly enhanced even in less extreme instances. For most of human history, the influence of any individual was limited. Actions were: (i) limited to proximity in place, (ii) limited to proximity in time, (iii) non-cumulative in extent, (iv) mostly the result of non-expert knowledge. These restrictions have now largely been removed and especially so for engineers in their professional activities. The examples of cluster munitions and water give some indication of the range of influence, which extends far into the very structure of society and into opportunities and standards in civil life. The extent of these influences brings with it great responsibilities. It is important that engineers are aware of these responsibilities and that they can articulate their professional priorities clearly. For example, it is vital that they can argue the case for freeing their skills from uninformed and unimaginative military and political control and rather apply their expertise for a long-lasting peace and prosperity in place of, at best, a transient peace based on fear and the anticipation of future war.

Peace is more than the absence of conflict. It is characterised by relationships between individuals, and social groupings of all sizes, based on honesty, fairness, openness and goodwill. However, the presence of violence within, and especially between, human societies shows that there is also a darker side to human nature. Thus, peace is always fragile and its maintenance requires the continued engagement of all. This is especially so when we take our places in larger groupings such as at our places of work or in our roles as citizens of nation states. For example, it is highly unlikely that many of the engineers professionally involved in the design or use of indiscriminate weapons would use unprovoked and indiscriminate extreme violence in their private, personal lives. It seems that we can have an unfortunate ability to bracket our ethical values in professional circumstances.

If engineering is to develop an aspiration of giving priority to helping people over technical ingenuity it would be of great benefit if means could be found to promote

[4] SIPRI (2007) estimates there are roughly 1700 tonnes of highly enriched uranium and 500 tonnes of separated plutonium in the world, sufficient to produce over 100,000 nuclear weapons.

a culture of peace *within* the profession. Such a culture would, for example, seek to redress imbalances such as those that have already been noted in the case of military technology and water. Further, it may be suggested that such a culture of peace within engineering could have the greatest likelihood of being established and provide the greatest benefits if it were aligned with the goals of other like-minded individuals and institutions. For example, the engineering profession may better enhance prospects of meeting its aspirations if its goals are aligned with those of major initiatives such as the UN activity for the promotion of a *Culture of Peace* (UN 2006). Such a culture is considered to consist of values, attitudes and actions that promote cooperation and mutuality among individuals, groups and nations. However, this initiative itself shows the lack of awareness of the capabilities of engineers. The initial United Nations *Declaration and Programme of Action on a Culture of Peace* recognised that key roles could be played by "parents, teachers, politicians, journalists, religious bodies and groups, intellectuals, those engaged in scientific, philosophical and creative and artistic activities, health and humanitarian workers, social workers, managers at various levels as well as non-governmental organisations" (UN 1999). It is notable that engineers are not mentioned in this list, nor are they mentioned in any of the primary documents related to a *Culture of Peace*. Maybe engineers are considered to be too closely aligned to the military. Maybe the omission was a simple error. Whatever the explanation, it is timely and necessary to remedy the omission.

For reasons of which those outlined in the present chapter are key examples, an analysis will now be developed that will provide a philosophical grounding for an aspirational ethical ethos for engineering. This opens in the next chapter with a consideration of the traditional ethical viewpoints that have been adopted by engineers.

References

Broers A (2005) The triumph of technology, Cambridge University Press, Cambridge

Cabinet Office (2006) The future of the United Kingdom's nuclear deterrent, London, HMSO

Gansler JS (2003) Integrating civilian and military industry, Issues in Science and Technology. http://www.issues.org/19.4/updated/gansler.html

Hutton G, Haller L (2004) Evaluation of the costs and benefits of water and sanitation improvements at the global level, World Health Organisation, Geneva

Jonas H (1984) The imperative of responsibility, Chicago University Press, Chicago. Originally published as Das Prinzip Verantwortung (1979) Insel Verlag, Frankfurt am Main

Jones RG (1998) Peace, violence and war. In: Christian Ethics, Hoose B (ed), Continuum, London, p. 210–222

Marx K (1845) Theses on Feuerbach, No. 11

McGrath R (2004) Cluster munitions – weapons of deadly convenience? Reviewing the legality and utility of cluster munitions, presentation at Meeting of Humanitarian Experts, Geneva, 11 November 2004

Stockholm International Peace Research Institute (2007) SIPRI Yearbook 2007, Armaments, disarmament and international security, Oxford University Press, Oxford

Stockholm International Peace Research Institute (2008) SIPRI Yearbook 2008, Armaments, disarmament and international security, Oxford University Press, Oxford
United Nations (1999) Declaration and programme of action on a Culture of Peace, General Assembly, A/RES/53/243, UN, New York
United Nations (2000) United Nations Millennium Declaration, General Assembly, A/RES/52/2, UN, New York
United Nations (2006) Culture of Peace, General Assembly, A/61/175, UN, New York
United Nations Educational, Science and Cultural Organization (2003) Water for people, water for life: World Water Development Report, UNESCO, Paris
United Nations Educational, Science and Cultural Organization (2006) Water: a shared responsibility: World Water Development Report 2, UNESCO, Paris
United Nations Foundation (2008) Conflict prevention and peace building, UNF, New York
United Nations News Centre (2006) Israel's immoral use of cluster bombs in Lebanon poses major threat – UN aid Chief, 30 August 2006, UN, New York

Chapter 3
Traditional Ethical Viewpoints

3.1 Introduction

As a minimum, an ethical outlook requires that our aims and actions are sensitive to the needs of others. Our everyday personal experiences indicate that, at least in some instances, people act so as to be genuinely helpful to each other rather than to further some ultimately selfish goal. However, in an age when the concept of the "selfish gene" has an underlying influence on our thinking about the world, there is a tendency to assume that such helpful and unselfish behaviour is in some way unnatural. This assumption needs attention before consideration of specific ethical viewpoints.

Contrary to such a tendency to view the selfish as natural, recent studies have emphasised that evolution strongly favours animals that assist each other (de Waal 2006; Joyce 2006). It seems likely that empathy first became explicitly apparent in the context of mammalian parental care. It is then possible to envisage a broadening of concern to close kin and further to the members of larger groups. Indirect reciprocal help may have played a significant role in this development. Individuals can have much to gain from a strong community. Even a level of concern for strangers can find an evolutionary explanation if community membership is fluid. With the development of language the formulation of evaluative moral concepts became possible so that actions could be thought not only beneficial but also good. The development of language and the possibility of self-evaluation were also most likely intimately related. A growing ability to reason and assess would have helped clarify the ways in which an individual could make a difference to a community and eventually lead to the expectation that an individual was expected to make a positive contribution.

A most important conclusion of these evolutionary studies is that helpful behaviour is an essential part of our constitution and not simply a veneer laid over our biological nature by culture or religion. Correspondingly, modern ethics should not be seen as a set of obstacles placed in the way of the pursuit of our interests but rather as the study and practice of ways of relating to other people that are natural and beneficial. For example, many will be able to recognise the benefits of parental care from their own experience. Conversely, there will generally be agreement that mindless violence is not helpful. An essential corollary of this view is that ethics permeates our lives rather than being an additional activity that we adopt when a dilemma arises.

Consideration of the evolutionary aspects of helpful behaviour indicates some development from an emotional type of response of the parent-child nature, through increasingly social types, to a more intellectual type of reason and assessment. In our present state of development these types remain closely entwined and they are, indeed, often synergistic. However, most published considerations of ethics have emphasised the importance of reasoned assessment. Such rational approaches have differed, among other things, according to the degree of objectivity considered possible in ethical assessments. It can be helpful to consider four possibilities (Graham 2004: 14):

Hard subjectivism – that for moral evaluations there are *never* any "right" answers.
Soft subjectivism – that for *many* moral evaluations there are no "right" answers.
Soft objectivism – that for any moral evaluation there *may* be a "right" answer.
Hard objectivism – that for *every* moral evaluation there is a "right" answer.

The present book will adopt the position of soft objectivism, for at the very least this allows a rational discussion to proceed. It should be noted, however, that it can be very difficult to define in any given context the meaning of "right", and even if this term could be clearly defined it could still be difficult to assess the relative merits of different moral options.

Life is complex, and ethics and moral decisions may correspondingly be expected to be complex. The most influential philosophical approaches to ethics can give an initial appearance of great clarity, which is a significant part of their wide appeal. In all cases, however, this initial clarity is accompanied by a failure to capture important features of the nature of human interactions. Ethical principles cannot be framed simply in a way that avoids equivocal cases or the need for exceptions. The more general the framing of the principles the more likely it becomes that such tricky cases or exceptions will arise. Furthermore, it is also often rather easier to identify actions that should be avoided rather than to recommend a specific action.

It is not the purpose of the present chapter to provide a systematic analysis of the different influential philosophical approaches to ethics. Rather, four approaches that have been influential in engineering will be outlined and their advantages and disadvantages when applied to engineering will be considered. The approaches to be considered are consequentialism, contractualism, duty and virtue.

3.2 Consequentialism

There is a tendency for engineers to adopt as a default position in the consideration of ethical problems a calculation of consequences and risks, a calculus of consequences. The philosophical basis of this approach to ethics is termed consequentialism, the fundamental nature of which may be specified as follows:

> ... postulate some good as the human good and then seek to identify the act that will maximise that good; that act is (by definition) the act of greatest utility and (by ethical stipulation) the right act. (Finnis 1983: 81)

Such an approach has a number of initially attractive features, especially for those with a scientific or engineering background. These features include simplicity and the quasi-empirical basis that the maximisation of a good suggests. In this context it should, however, be noted that a normative ethical approach of this type starts from the postulation of a general principle and leads to proposed action in specific instances, whereas scientific empiricism begins with specific observations which may eventually lead to a general theory. Part of the attraction for engineers is also a similarity to the familiar and more limited exercises of cost-benefit analysis.

Utilitarianism is the most widely discussed form of consequentialism and the most enduring account of this approach is Mill's *Utilitarianism* which:

> ... accepts as the foundation of morals "utility" or the "greatest happiness principle" ... that actions are right in proportion as they tend to promote happiness; wrong as they tend to produce the reverse of happiness. By happiness is intended pleasure and the absence of pain; by unhappiness, pain and the privation of pleasure ... that pleasure and freedom from pain are the only things desirable as ends ... that standard is not the agent's own greatest happiness, but the greatest amount of happiness altogether. (Mill 1861/2001: 7, 11)

Reference to happiness has an immediate appeal for, as in the case of parental-care, few would object to it. Further, as ethics is a practical pursuit, all ethical approaches should be sensitive to the outcomes of proposed actions. However, utilitarianism is a key example of how initial apparent clarity can lead to a clear failure to capture important features of human interactions on application of the guiding principle. Mill himself wrote about many of the weaknesses and a vast literature has ensued (recent insightful analyses have been given by Smart and Williams (1973), Mackie (1977), and Finnis (1983)). Among the difficulties are:

- Few would limit the end of human activities solely to happiness or pleasure (which are equated by Mill).
- The implied calculus of pleasure and pain could only be carried out in an arbitrary manner, and calculations become even less viable if a more realistic range of the various ends of human activities is taken into account.

- The outcomes of any action depend on a number of uncertain factors, including the actions of others, and are theoretically infinite in extent including both intended and unintended consequences and consequences arising because another action was not chosen.
- The principle of action is impersonal – it answers the question, "What should be done?" rather than, "What should *I* do?". It cannot capture the human concern for family and friends.
- In principle any action is allowed.
- There is no provision for justice – there is no provision regarding the distribution of the proposed good.

The neglect of justice was recognised by Mill as one of the strongest obstacles to his approach (Mill 1861/2001: Chap. 5). The possibility that utilitarianism could allow any action including "the judicial condemnation of the innocent" (Anscombe 1958/1997: 44) has been a major criticism of many modern commentators. The impersonal nature of such a proposed assessment has led to doubts as to whether utilitarianism can be described as an *ethical* system at all.

Advocates of utilitarianism have provided sophisticated analyses that seek to address criticisms. A significant reason for the endurance of Mill's account is that it anticipates the basis of many of the key features of these later analyses. He recognised different levels of good with the famous statement, "It is better to be a human being dissatisfied than a pig satisfied; better to be Socrates dissatisfied than a fool satisfied" (Mill 1861/2001: 10). He recognised that the worth of an agent should be distinguished from the morality of an action and that the interests of persons were complex (Mill 1861/2001: 18, 20). He recognised the need for rules or subordinate principles to avoid the necessity of a complete calculation of consequences for each action (Mill 1861/2001: 23–25). These comments have led to an extended debate about rule-utilitarianism as contrasted to act-utilitarianism. There are great practical advantages in having predetermined rules for action. However, there is a danger that adherence to the rules becomes the principle of action, and it has also been argued that the need to allow for exceptions to rules means in practice that rule-utilitarianism collapses into act-utilitarianism.

Even a brief discussion of consequentialism is enough to present serious doubts about any analysis of ethics in which everything depends on the consequences. However, putting aside these doubts for a moment, it is interesting to consider briefly how a consequentialist analysis would evaluate the prioritisation of the activities considered in Chap. 2. Justification of the human devastation caused by cluster weapons or nuclear weapons would require a correspondingly huge range of human benefit if any calculus were to provide a favourable outcome. In the case of drinking water and sanitation provision to the poorest communities, it should be observed that on a consequentialist basis the removal of such a lack would greatly increase human wellbeing of those concerned with little expected negative impact on others. However, we put enormous engineering resources into weapons development and relatively little into drinking water and sanitation. The most likely consequentialist justification for this strange prioritisation is that we value the wellbeing of certain groups far higher than those of others. In the words of Mill, "there are large portions of mankind whose happiness it is still practicable to disregard" (1861/2001: 33). Mill hoped that this regrettable state of affairs might be ameliorated "by the contagion of sympathy and the influences of education", effects for which we are still waiting.

The view that everything in ethics depends on the consequences is inadequate. However, the view that consequences are important is essential to ethics, especially if matched by the contagion of sympathy in a more thorough-going approach.

3.3 Contractualism

Contractualism regards the motivation for acting ethically as being social agreement backed by a regulative framework. A celebrated early version of such a contract is developed by Socrates in Plato's dialogue *Crito*, where it is said to be established between the laws or the city and each citizen (Plato c.395 BC/1997). The approach has been refined in many ways since, with the most significant modern work being Rawls' *A Theory of Justice* which claims to provide, "an alternative systematic account of justice that is superior ... to the dominant utilitarianism of the tradition" (Rawls 1971/1999)[1]. Rawls' particular aim is to present a theory of justice as a viable systematic doctrine so that the maximising of good does not maintain dominance by default. The presentation proposes a hypothetical contract between individuals as the basis for institutional arrangements up to the scale of nation states.

A Theory of Justice is a large and complex work, though its starting point is a concept of apparently great simplicity. This is a situation in which people who are to live together choose by agreement the principles that are to be used to assign basic rights and duties and to determine the division of social benefits. The key feature is that this agreement is to be reached behind a "veil of ignorance". The veil of ignorance ensures that no one knows his or her social status or wealth, or his or her fortune in the distribution of intelligence, health and the like. It is even assumed that they do not know their conception of the good or their particular psychological propensities. This is a rather curious state of affairs as such a lack of specific knowledge must be accompanied by sufficient general knowledge regarding the operation of a society for the discussion to proceed – it is a hypothetical situation. Very briefly stated, Rawls concludes that people in this initial situation would choose a combination of two rather different principles:

[1] "The tradition" refers to the Western philosophical tradition.

- The first is related to liberty: each person is to have a right to the most extensive total system of equal basic liberties compatible with a similar right for all.
- The second is related to justice as fairness: equality of opportunity for all and a limit on the gap that is allowed to arise between the advantaged and disadvantaged, hence encouraging the willing cooperation of everyone.

Rawls' work has generated a vast literature, perhaps as with utilitarianism because of the simplicity and scope of the starting point. Likewise, it is undoubtedly a constructive and useful approach. It is also rather demanding, as the hypothetical situation requires a group of highly intelligent and rational people (one might also add Western and liberal) prepared to discuss issues of great complexity. The approach assumes that the members of the group are rather reluctant to take risks. It can also be criticised for giving insufficient explicit attention to the worth and dignity of individual persons. Indeed, no specific ethical values appear explicitly in its formulation.

However, again putting aside doubts for a moment, it is interesting to consider briefly how a contractualist analysis like that of Rawls would evaluate the prioritisation of the activities considered in Chap. 2. It seems highly unlikely, assuming a "veil of ignorance", that a hypothetical group including all affected individuals would agree a prioritisation that led to the spending of enormous sums on military equipment while at the same time neglecting to invest a much smaller sum so that the basic needs for drinking water and sanitation for billions of the most disadvantaged could be met. Such prioritisation violates both the principle of equal basic liberties and the principle of justice as fairness. It is also unlikely that a decision to develop weapons of high devastation potential would be agreed even if these basic needs were met, not only because of the ignorance of the members about their individual situation but also because they are intelligent and risk-averse, and would hence realise that even accidental or "limited" use could be catastrophic. The currently existing prioritisation, and many other instances of unjust distribution of benefits, exist to a large extent because of a basic weakness in agreement backed by regulative frameworks – that the most disadvantaged are excluded from the formulation of the agreement.

A further weakness of contractual approaches shown by the analysis in Chap. 2 for the case of weapons is that even where there is widespread formal assent to international agreements, such as the case with *Protocol 1 Additional to the Geneva Conventions*, such agreements are not adhered to in practice. Many countries carry out breaches. Such breaches are, however, only possible with weapons designed, manufactured and deployed using the skills of engineers. Just how far international agreements are neglected in practice can be illustrated by comparing our knowledge of the current world with the *Preamble to the Charter of the United Nations* (UN 1945) that now has 192 parties:

WE THE PEOPLES OF THE UNITED NATIONS DETERMINED
to save succeeding generations from the scourge of war, which twice in our lifetime has brought untold sorrow to mankind, and
to reaffirm faith in fundamental human rights, in the dignity and worth of the human person, in the equal rights of men and women and of nations large and small, and
to establish conditions under which justice and respect for the obligations arising from treaties and other sources of international law can be maintained, and
to promote social progress and better standards of life in larger freedom,

AND FOR THESE ENDS
to practice tolerance and live together in peace with one another as good neighbours, and
to unite our strength to maintain international peace and security, and
to ensure, by the acceptance of principles and the institution of methods, that armed force shall not be used, save in the common interest, and
to employ international machinery for the promotion of the economic and social advancement of all peoples,

HAVE RESOLVED TO COMBINE OUR EFFORTS TO ACCOMPLISH THESE AIMS.

Additionally, with respect to the context of the present book, if contractualism in itself really were a convincing approach, then it might be expected that engineers would be influenced in the prioritisation of the use of their skills by international agreements, such as the *United Nations Charter*, *Protocol 1 Additional to the Geneva Conventions* or the *Treaty on the Non-proliferation of Nuclear Weapons*. For many individuals this appears not to be so. It could be argued that it is too much to expect an individual to be aware of and to be influenced in their behaviour by such major international agreements, which might be considered to lie in the realm of international politics rather than ethics. The case against this viewpoint will be presented in a later chapter. Here, a type of contractualism that applies more directly to engineers will be considered.

Practising engineers are very often corporate members of recognised professional institutions (for chemical engineering, civil engineering, mechanical engineering and so forth). Such engineering institutions have the duty to regulate the conduct of their members and to apply appropriate disciplinary procedures in the case of breaches of good practice. In the UK, engineering institutions are licensed by the Engineering Council which also aims to "set and maintain realistic and internationally recognised standards of professional competence and ethics for engineers". Competence is recognised by the awarding of Chartered Engineer status. With respect to ethics, the Engineering Council expects each licensed engineering institution to develop a Code of Professional Conduct "placing a personal obligation on its members to act with integrity, in the public interest and to exercise all reasonable professional care". The guidelines for these codes include the following (ECUK 2007):

1. Prevent avoidable danger to health or safety.
2. Prevent avoidable adverse impact on the environment.

3. a Maintain their competence.
   b Undertake only professional tasks for which they are competent.
   c Disclose relevant limitations of competence.
4. a Accept appropriate responsibility for work carried out under their supervision.
   b Treat all persons fairly, without bias, and with respect.
   c Encourage others to advance their learning and competence.
5. a Avoid where possible real or perceived conflict of interest.
   b Advise affected parties when such conflicts arise.
6. Observe the proper duties of confidentiality owed to appropriate parties.
7. Reject bribery.
8. Assess relevant risks and liability, and if appropriate hold professional indemnity insurance.
9. Notify the Institution if convicted of a criminal offence or upon becoming bankrupt or disqualified as a Company Director.
10. Notify the Institution of any significant violation of the Institution's Code of Conduct by another member.

These guidelines give a clear statement of the behaviour expected of an engineer when engaged in a specified professional activity. However, they provide the basis for a set of moral norms rather than providing the basis for a real ethics of engineering. They do not encourage a genuinely *aspirational* engineering ethical ethos. They give, for example, no indication of how an engineer should aspire to make the best use of his or her skills. Maybe that was never the intention. However, reading something like these guidelines is as close as many engineers come to encouragement to make the most effective use of their skills.

Overall, contractualism by itself fails to promote an ethos of high ethical aspirations at (at least) two levels. At one level, the aspirations of ambitious international agreements are to a large extent ignored in practice even though they are formally assented to by most countries. At the second level, more specific professional codes, to which individuals are explicitly required to assent, tend to be limited to securing an acceptable functioning of existing activities.

3.4 Kantian Duty

A further view that has been influential in discussions of engineering ethics is based around the idea of duty, an approach that was given a succinct formulation in Kant's *Groundwork of the Metaphysics of Ethics*. Kant's objective was highly ambitious. He sought to formulate and establish, from first principles and on the basis of reason alone, the "supreme principle of morality" (Kant 1785/1998: 5). He thought that moral concepts established in this way would provide an obligation, a duty, to act accordingly, "for the sake of the law" (Kant 1785/1998: 3)[2].

---

[2] Law in the sense of "moral law".

Kant formulated his ethics in a number of ways of which two are especially relevant in the present context. He proposed that there was one "categorical imperative", that is a rule which should be followed whatever the circumstances:

> Act only in accordance with that maxim through which you can at the same time will that it becomes a universal law. (Kant 1785/1998: 31)

He further proposed that every person existed as an end in himself or herself, leading to the formulation of a "practical imperative", a strong statement of respect for persons:

> So act that you use humanity, whether in your own person or in the person of any other, always at the same time as an end, never merely as a means. (Kant 1785/1998: 38)

The categorical imperative is useful mostly for ruling out actions rather than for selecting positively for favoured actions, for several options may in principle meet the criterion of universality. Kant further wished to emphasise the positive aspects of the practical imperative, so he proposed that everyone should try as far as possible to further the ends of others, a "kingdom of ends", a "systematic union of various rational beings through common laws" (1785/1998: 41). There is here the implication of reason leading to rational agreement and hence to an obligation to follow the resulting laws.

This brief description gives rise to several observations in the present context. Firstly, any scientist or engineer is likely to be wary of a claim to have discovered a supreme principle without any empirical input. Kant aims to rule out the validity of such input, and specifically excludes reference to happiness, "it is a misfortune that the concept of happiness is such an indeterminate concept ... without exception empirical" (1785/1998: 28). Secondly, it appears that the motivation for action is rather dry, obedience to duty. Thus, even though Kant's statement of respect for persons is very strong, it does not really allow a motivation based on personal compassion. Kant himself recognised this weakness and wrote about the need for "a feeling of pleasure or of delight" if the rational prescription of the "ought" of duty is to lead to action (1785/1998: 64).

Kant's emphases on duty, intention, respect for persons and impersonal cooperation have been very influential in philosophical ethics often leading to dense argumentation (Korsgaard 1996). These emphases have also played a significant role in engineering ethics, especially the concept of duty as a background to the formulation of ethical codes. Kant's argumentation leading to the formulation of the categorical and practical imperatives can be difficult to follow, but the resulting statements are certainly aspirational in outlook. However, it appears that they have not had much influence on the prioritisation of weapons and water considered in Chap. 2. The universal use of highly devastating weapons would be disastrous; the lack of provision for drinking water and sanitation for large numbers of the world's population, even though we have the means to do so, hardly suggests an unbiased respect for persons.

3.5 Virtue Ethics

Virtue ethics has previously had less influence on engineering ethics than consequentialism, contractualism or duty ethics. However, it merits consideration here as its viewpoint has had an underlying influence on philosophical ethics and it provides important elements of a basis for the development of an aspirational engineering ethic. As for utilitarianism and duty ethics, the influence of virtue ethics benefits greatly from the existence of a succinct foundational text, in this case Aristotle's *Nichomachean Ethics* (NE) (350 BC/1998), which 2350 years later than its origins remains essential reading for serious students of ethics.

In contrast to many later writers, Aristotle was clear from the outset that ethics was a very difficult subject, that human goods were exceedingly complex, and hence that modesty was required in ethical analysis, "We must be content, then, in speaking of such subjects and with such premises to indicate the truth roughly and in outline ..." (NE 1094^{b}19). He was greatly concerned with how man could lead a complete, fulfilled life and especially sought to identify the overriding "function" or essential activity of man (which could in part be flute-player, sculptor, artist and so on). The two essential features of such a function he identified as "practical wisdom" and "moral excellence", "Again, the function of man is achieved only in accordance with practical wisdom as well as with moral excellence: for excellence makes the aim right, and practical wisdom the things leading to it" (NE 1144^{a}7).

Aristotle's emphasis was always on the practical nature of ethics, not in knowing what is good but in being and doing good. Wisdom he regarded as knowledge, of the most fundamental type possible, combined with comprehension. Practical wisdom was, however, not just concerned with principles but rather especially with particulars. It was the ability to recognise the things about which it was possible to productively deliberate, and further, "a reasoned and true state of capacity to act with regard to human goods" (NE 1140^{b}20). Experience was an important aspect of practical wisdom.

However, underlying practical wisdom was moral excellence, that is virtue. Aristotle was clear that the identification and promotion of virtue were themselves complex tasks. He recognised that virtues were very often in an intermediate position between two corresponding vices and that it could hence be difficult to find the balance necessary to achieve the mean, "to do this to the right person, to the right extent, at the right time, with the right aim, and in the right way" (NE 1109^{a}27). As examples of virtues Aristotle considered, courage, temperance, liberality, magnificence, pride (between vanity and excessive humbleness), honour, good temper, tact and a number without name as only the "extremes" were specified.

Aristotle identified the greatest virtues as justice and friendship. Regarding justice, "it is complete excellence its fullest sense, because it is the actual exercise of complete excellence. It is complete because he who possesses it can exercise his excellence towards others too and not merely by himself" (NE 1129^b30). He considered justice to be closely related to friendship. Aristotle identified three types of friendship, for utility, for pleasure and a perfect friendship based on moral excellence. He recognised that the latter type of friendship alone endured, though it depended on community and a certain equality between the friends. For Aristotle, such friendship was certainly an excellence promoting high aspirations, "For friendship asks a man to do what he can, not what is proportional to the merits of the case ... to the utmost" (NE 1163^b15).

There is considerable sophistication in *Nichomachean Ethics* compared to the foundational texts of the other ethical viewpoints. Thus, Aristotle recognised an interpretation of happiness as pleasure as a platitude, and the Greek word *eudaimonia,* which is often translated into English as "happiness", has a much more active meaning, rather "living and faring well" or flourishing. He noted that there are many things we value even if they bring no pleasure. He recognised that some actions may not be good however they are carried out, such as "committing adultery with the right woman, at the right time, and in the right way" (NE 1107^a15). He recognised the importance of intention, "to do what is unjust is not the same as to act unjustly" (NE 1136^a26). Indeed, he identified three conditions for a person's acts to be just, "The agent must also be in a certain condition when he does them; in the first place he must have knowledge, secondly he must choose the acts, and choose them for their own sakes, and thirdly his action must proceed from a firm and unchangeable character" (NE 1105^a30). Aristotle also recognised that universal statements, supreme principles, have limitations, "For among statements about conduct those which are general apply more widely, but those which are particular are more true, since conduct has to do with individual cases, and our statements must harmonise with the facts in these cases" (NE 1107^a29).

Even though Aristotle lived in a relatively small community, a city-state, he realised that friendship with very many was not possible. In its place he proposed "goodwill", as an extension of the term, a type of friendly relation though lacking intimacy. This is an important observation, now exceedingly relevant in the context of the reach of relationships in the modern world. Regarding the modern issues raised in Chap. 2, Aristotle's guiding virtues of justice and friendship are highly incompatible with massively devastating weapons. In the vocabulary of *Nichomachean Ethics* the use of such weapons would have been described as "vicious" rather than virtuous. He would also have recognised the plight of those without drinking water and sanitation, for he was fully aware of the role of material benefits in enabling human flourishing.

The present chapter opened with reference to some very recent animal studies showing the naturalness of helpful behaviour. It is remarkable that Aristotle considered such matters, commenting that "some even of the lower animals have practical wisdom, viz. those which are found to have a power of foresight with regard to their own life", also noting, "they have no universal beliefs but only imagination and memory of particulars" (NE $1141^{a}27$, $1147^{b}3$). Some writers have recently argued in a more philosophical way for continuity in concepts of goodness promoting flourishing in all forms of life, from plants, to non-human animals, to humans. In such an analysis, moral evaluation of human actions and dispositions is seen as one example of a genre of evaluation common to all living things. For example, vice can be seen as a form of natural defect and virtue as goodness of the will (Foot 2001).

In the overall context of an account of engineering ethics and virtue ethics, it is interesting to take note of Benjamin Franklin, one of the greatest innovators of the eighteenth century. Franklin's many inventions include lightning conductors, based on his studies of electricity, and the Franklin stove, which produced more heat and less smoke than an ordinary fireplace. He also made quantitative studies of refrigeration by evaporation. He was a pioneer of biomedical engineering – now one of the cutting-edge fields of engineering – inventing bifocal glasses, a flexible urinary catheter and an extension device allowing people with arthritis to handle small objects. His investigations into the use of electricity in medicine were probably literally hair-raising, but ultimately unsuccessful. He was also very concerned with public safety, forming the Union Fire Company, the first organisation of its type in America to be organised to tackle house fires, equipped with, "fire-engines, ladders, fire-hooks, and other useful implements" (Franklin 1788/1996: 82). Franklin was very likely the first engineer to guide his activities explicitly on the basis of virtue ethics. As a young man he specified the following virtues, intending to acquire their "habitude": temperance, silence, order, resolution, frugality, industry, sincerity, justice, moderation, cleanliness, tranquillity, chastity, humility (Franklin 1788/1996: 64). His aspiration to help others found explicit expression of virtue in his refusal to patent his innovations, "That, as we enjoy great advantages from the inventions of others, we should be glad of an opportunity to serve others by any invention of ours, and this we do freely and generously" (Franklin 1788/1996: 92).

Franklin was able to make important original contributions in the diverse fields of engineering, medicine and business. In all his activities he was strongly motivated by an aspirational ethical ethos. It is rare for an individual to combine the breadth of knowledge necessary for such achievement. However, the development of an aspirational ethical ethos for today's engineers can benefit from knowledge of ethical approaches in medicine and business. This is the subject of the next chapter.

References

Aristotle (1998) Nichomachean ethics. In: The complete works of Aristotle, vol 2, Barnes J (ed), Princeton University Press, Princeton. Original attribution 350 BC
Anscombe GE (1997) Modern moral philosophy, in "Virtue ethics", Crisp R and Slote M (eds), Oxford University Press, Oxford. Originally published 1958
de Waal F (2006) Primates and philosophers, Princeton University Press, Princeton
Engineering Council UK (2007) Guidelines for Institution codes of conduct, ECUK, London
Finnis J (1983) Fundamentals of ethics, Clarendon Press, Oxford
Foot P (2001) Natural goodness, Oxford University Press, Oxford
Franklin B (1996) The autobiography, Dover Publications, New York. Originally published 1788
Graham G (2004) Eight theories of ethics, Routledge, London
Joyce R (2006) The evolution of morality, MIT Press, Cambridge
Kant I (1998) Groundwork of the metaphysics of morals, Cambridge University Press, Cambridge. Originally published as Grundlegung zur Metaphysik der Sitten, 1785
Korsgaard CM (1996) The sources of normativity, Cambridge University Press, Cambridge
Mackie JM (1977) Ethics – inventing right and wrong, Penguin Books, London
Mill JS (2001) Utilitarianism, Hackett Publishing Company, Indianapolis. First published 1861
Plato (1997) Crito. In: Plato: Complete Works, Cooper JM (ed) Hackett Publishing Company, Indianapolis. Original attribution c. 395 BC
Rawls J (1999) A theory of justice, revised edition, Oxford University Press, Oxford. First published 1971
Smart JJC, Williams B (1973) Utilitarianism: for and against, Cambridge University Press, Cambridge
United Nations (1945) Charter of the United Nations, UN, New York

Chapter 4
Ethics in Medicine and Business

As noted in Chap. 1, Aristotle refers in the opening section of *Nichomachean Ethics* not only to engineering (shipbuilding), but also to medical art and economics. A consideration of ethical practices in the corresponding modern professions of medicine and business can be beneficial in the development of an aspirational engineering ethos.

4.1 Ethics in Medicine

Medicine is the profession that has had the most intense interest in ethical concerns. Issues in medical ethics are debated vigorously both within the profession and in the public sphere. Recently, the boundaries of medicine have been increasingly difficult to define. This has led to the stimulation of ethical debate in areas of overlap with other disciplines, especially the biosciences. Certain aspects of engineering are similarly brought into consideration through the rapid growth in biomedical engineering.

Medicine has provided the most ancient and most widely known code of professional ethics, the Hippocratic Oath. The exact origins of the oath are somewhat uncertain, but it is thought to date from about the 4th century BC. The oath was sworn "by Apollo the physician, and Asclepius, and Hygieia and Panacea and all the gods and goddesses as my witnesses", a feature which, with some other culturally determined characteristics, does not lead to its complete commendation in modern times. However, many of its features are concerned with issues that remain current, including promises not to cause abortion or euthanasia, confidentiality, working within one's competence and avoidance of improper personal relationships. The oath also contains the words "and I will do no harm or injustice to them", a form of which is often seen as leading to a key axiom of medical practice, "First, do no harm".

There is currently a great diversity of medical ethical codes. Amongst the most authoritative are those produced by the World Medical Association (WMA), a confederation of the representative medical associations of 84 countries (in the case of UK, the British Medical Association). The WMA's main goal is to establish and promote the highest possible standards of ethical behaviour and care by doctors. The WMA has a number of agreed policy statements on aspects of medical ethics. For example, the *Declaration of Geneva*, first adopted in 1948 and most recently revised in 2006, could be seen as a modern version of the Hippocratic Oath (WMA 2006a):

> AT THE TIME OF BEING ADMITTED AS A MEMBER OF THE MEDICAL PROFESSION:
> I SOLEMNLY PLEDGE to consecrate my life to the service of humanity;
> I WILL GIVE to my teachers the respect and gratitude that is their due;
> I WILL PRACTISE my profession with conscience and dignity;
> THE HEALTH OF MY PATIENT will be my first consideration;
> I WILL RESPECT the secrets that are confided in me, even after the patient has died;
> I WILL MAINTAIN by all the means in my power, the honour and the noble traditions of the medical profession;
> MY COLLEAGUES will be my sisters and brothers;
> I WILL NOT PERMIT considerations of age, disease or disability, creed, ethnic origin, gender, nationality, political affiliation, race, sexual orientation, social standing or any other factor to intervene between my duty and my patient;
> I WILL MAINTAIN the utmost respect for human life;
> I WILL NOT USE my medical knowledge to violate human rights and civil liberties, even under threat;
> I MAKE THESE PROMISES solemnly, freely and upon my honour.

This is a highly aspirational statement written in language that aims to be culturally neutral.

One of the significant differences between medicine and engineering as professions is that in medicine there are national bodies with strong legal sanctions whose purpose is to protect the public by ensuring proper standards of medical practice. In the UK this role is fulfilled by the General Medical Council (GMC) with which doctors must be registered to practise. The main functions of the GMC are defined by law to be: keeping up-to-date registers of qualified doctors, fostering good medical practice, promoting high standards of medical education and dealing firmly and fairly with doctors whose fitness to practise is in doubt. As a contribution to these ends, the GMC has published *Good Medical Practice*. This opens with a statement of the duties of a registered doctor (GMC 2006):

> Patients must be able to trust doctors with their lives and health. To justify that trust you must show respect for human life and you must:
> Make the care of your patient your first concern
> Protect and promote the health of patients and the public
> Provide a good standard of practice and care

– Keep your professional knowledge and skills up to date
– Recognise and work within the limits of your competence
– Work with colleagues in the ways that best serve patients' interests

Treat patients as individuals and respect their dignity

– Treat patients politely and considerately
– Respect patients' right to confidentiality

Work in partnership with patients

– Listen to patients and respond to their concerns and preferences
– Give patients the information they want or need in a way they can understand
– Respect patients' right to reach decisions with you about their treatment and care
– Support patients in caring for themselves to improve and maintain their health

Be honest and open and act with integrity

– Act without delay if you have good reason to believe that you or a colleague may be putting patients at risk
– Never discriminate unfairly against patients or colleagues
– Never abuse your patients' trust in you or the public's trust in the profession.

You are personally accountable for your professional practice and must always be prepared to justify your decisions and actions.

This opening statement and the text of *Good Medical Practice* provide guidance rather than a statutory code, but it is made clear that serious or persistent failure to follow the guidance will put a doctor's registration at risk. The use of "must" in the opening statement indicates an overriding duty or principle. The term "should" is used later in the text to indicate an explanation of how to meet an overriding duty, or else where a duty or principle may not always apply.

This GMC statement has many features comparable to engineering codes of ethics such as Engineering Council UK guidelines for codes considered in Chap. 3 and the UK Royal Academy of Engineering *Statement of Ethical Principles* to be considered in Chap. 6. There are also two striking differences. Firstly, the emphasis on the *individual*, the patient, affected by the doctor's actions[1]. Engineering ethical codes are more impersonal in tone. Secondly, the final statement that the doctor is always *personally* accountable for decisions and actions tends not to receive the same explicit emphasis in engineering. These issues will be addressed in the second part of the present book.

In the context of the influence of institutions on ethics, the role of the World Health Organisation (WHO, a UN agency) should be considered. WHO regards *equity* in access to life-saving or health-promoting measures as being the global aim of health care. The organisation has a global role in promoting health care development and provision. Thus, another contrast between health-care and engineering is that whereas there is a single international agency playing an effective global role in health-care there are no comparable global organisations for engineering activities, with the exception of the International Atomic Energy Agency. This applies even in the case of water management, though WHO was one of the

[1] In practice, there is a growing emphasis on allowing the individual patient to make decisions about his or her treatment on the basis of the doctor's advice. I thank Dr Sunniva Eie Bowen for this information and other advice on practical aspects of modern medicine.

agencies responsible for drawing attention to the consequences of the inadequate provision of drinking water and sanitation in poor countries.

Among the policy statements of the WMA is a *Statement on Weapons of Warfare and Their Relation to Life and Health*, initially adopted in 1996 and revised in 2006 (WMA 2006b). This recognises that medical knowledge can be used in the development of new weapons systems targeted at specific human functions, such as vision, or at specific individuals or populations on a genetic basis. It declares that "The WMA believes that the development, manufacture and sale of weapons for use against human beings are abhorrent" and calls for a number of measures to support the prevention and reduction of weapons injuries. In the UK, the BMA has provided very explicit guidance:

> While doctors may have a legitimate role in reviewing the defensive capability of weapons, the BMA considers that doctors should not knowingly use their skills and knowledge for weapons' development. It objects to doctors' participation in weapons' development for the same reasons that it opposes doctors' involvement in the design and manufacture of torture weapons and more effective methods of execution: through their participation doctors are lending weapons a legitimacy and acceptability that they do not warrant. Doctors may consider that they are, in fact, reducing human misery through their involvement, but in reality the proliferation of weapons show this to be untrue.
> (BMA 2001: paragraph 37)

The response to an enquiry to the GMC as to whether the work of a doctor directly involved in weapons development would be compatible with continued GMC registration drew direct attention to this paragraph as a guidance statement. The GMC would make such decisions on a case by case basis with serious or persistent failure to follow the guidance of *Good Medical Practice* calling a doctor's registration into question. The response further stated, "paragraph 57 of *Good Medical Practice* says, 'you must make sure that your conduct at all times justifies your patients' trust in you and the public's trust in the profession'. Therefore, direct involvement of a doctor in the development of weapons systems which were targeting specific human functions or specific human groups might constitute conduct in breach of this paragraph, if it was considered to be serious enough"[2].

The last quoted sentence of this response refers to the growing interest in the development and use of drugs as weapons. The increasing militarization of biological knowledge facilitates the use of drugs as weapons to control emotions, memories or immune responses. Acquisition of genome and proteome information may allow the design of ethnically or even individually targetable pharmaceutical weapons. The development and deployment of such weapons require medical expertise. The BMA has carried out a detailed analysis of this issue, providing very strongly worded opposition, for example, "The BMA is fundamentally op-

[2] Enquiry by WRB, 25 May 2007. Response by Ms Farkhanda Maqbool, Policy Officer, GMC Standards and Ethics Team, 1 June 2007.

posed to the use of any pharmaceutical agent as a weapon", "the BMA does not believe it is part of doctors' role to develop weapons to harm people, even in order to fight terrorism, since that is contrary to the ethos of medical training", and "the duty to avoid harm rises above, for instance, a duty to contribute to national security" (BMA 2007a: v, 8, 20). The BMA emphasises instead that the upholding of international humanitarian law and human rights law is the best means of securing international security. In the present context, it is noteworthy that the manufacture and use of such pharmaceuticals requires extensive application of engineering, especially involving the skills of chemical and biochemical engineers. The strong statements of the medical profession on this issue bring the participation of engineers in such manufacture into a sharp ethical focus.

The practice of modern medicine would be rather limited without engineering contributions such as hospital design and construction, advanced diagnostic instrumentation, life-sustaining equipment and pharmaceutical manufacture. Further, in 2007 the *British Medical Journal* published a supplement entitled, *Medical Milestones. Celebrating Key Advances Since 1840*. The benefits of fifteen advances were highlighted: anaesthesia, antibiotics, chloropromazine, DNA structure, evidence-based medicine, germ theory, immunology, medical imaging, oral contraceptive pill, oral rehydration therapy, sanitation, risks of smoking, tissue culture and vaccines. In a subsequent poll, readers were invited to vote for the advance that they rated most highly. The most supported advance was sanitation (BMA 2007b). This recognition of the contribution of engineering to health is especially notable as the development of water supply and sewage disposal is not strictly speaking a *medical* advance. Regrettably, as detailed in Chap. 2, lack of such basic needs is still a major cause of illness and death in much of the world.

4.2 Philosophical Medical Ethics

Philosophical ethics has been widely applied to medicine. One of the best known approaches is developed in *Principles of Biomedical Ethics* by Beauchamp and Childress (2001). This develops a complex multi-layered approach. Four basic principles are proposed: respect for autonomy, nonmaleficence, beneficence and justice. These are in accord with the WMA and GMC statements. Autonomy refers to the respect due to the integrity of the individual patient. The authors argue that whereas nonmaleficence should always be observed there are more constraints on beneficence. They propose the usefulness of a distinction between specific beneficence, a duty to intimate others, and general beneficence to strangers. The latter leads to a kind of constrained utilitarianism. They view quantitative measures of beneficence such as cost-effectiveness analysis, cost-benefit analysis and risk-benefit analysis as useful aids to assessing overall utility, but regard autonomy and justice as setting limits on uses of these techniques. Justice is regarded as referring to the ways of distributing the benefits, risks and costs of health care fairly (apparently on a national basis). They discuss the complexity of the issues involved and propose an enforceable right to a decent minimum standard of healthcare within an allocation framework based on utilitarian and egalitarian standards.

Beauchamp and Childress emphasise the importance of setting goals beyond the moral minimum, of having high aspirations, such as occurs in an Aristotelian approach. They identify the five focal virtues for a doctor as being compassion, discernment, trustworthiness, integrity and conscientiousness. They also consider the four principles noted above as being virtues together with truthfulness and faithfulness. In this context they further observe that some very important aspects of a doctor's work, such as encouraging and cheering patients, are not included in professional codes.

In a part of their work that deals more extensively with an evaluation of formal ethical theories, Beauchamp and Childress propose eight criteria of adequacy: clarity, coherence, completeness and comprehensiveness, simplicity, explanatory power, justificatory power, output power and practicability. They then make an assessment of consequentialism, Kantian duty, rights-based theory, communitarianism and an ethics of care. In their assessment, all of these approaches are considered to have both benefits and drawbacks when applied to medical ethics. They regard the theoretical distinctions as being less important for practical ethics than philosophers often claim. As each approach has something useful to offer, they recommend combining the (for them) acceptable features of the different theories.

In their conclusions, Beauchamp and Childress contrast top-down approaches that emphasise general norms and ethical theory with bottom-up approaches that emphasise moral tradition, experience and particular circumstances. As neither general principles nor paradigm cases have sufficient power to generate conclusions with the reliability needed by medical ethics, they propose instead an integrated approach, a "coherence theory", that combines top-down and bottom-up approaches in a reflective equilibrium, a term they draw from Rawls (1971/1999). A coherence approach will be familiar to engineers, as the development of engineering understanding is based on an analogous interplay between empirical data and theory.

Such a coherence approach to justification is combined with moral values from what the authors term "common-morality" – "that we respect persons, take account of their well being, treat them fairly, and the like" (Beauchamp and Childress 2001: 401). The authors here seem to be referring especially to the core, underlying basis of morality, for they note that not all "customary" moralities qualify as part of the common-morality. This common-morality appears to be in accord with the conclusions of the recent studies on the evolution and naturalness of morality by de Waal, Foot and Joyce referred to in Chap. 3 of the present book, though Beauchamp and Childress are themselves strongly influenced by the work of W.D. Ross. Common-morality can be difficult to define precisely, though they could consider this an advantage, comparable to a certain degree of uncertainty in the definition of principles allowing negotiation and compromise on difficult issues. The authors' ultimate goal is to produce a type of medical ethics where, as is the case for what they consider "everyday" moral reasoning, "we effortlessly blend appeals to principles, rules, rights, virtues, passions, analogies, paradigms, narratives and parables" (Beauchamp and Childress 2001: 408)[3]. Those who propose a stronger role for principles in ethics have, of course, challenged this view. At the very least, the inclusion of the word "effortlessly" is rather optimistic!

Several features of Beauchamp and Childress's analysis merit consideration in the development of engineering ethics. Most important is their emphasis on the need for an aspirational approach as a counterbalance to what have been the prevailing viewpoints in philosophical ethics. It might also be remarked that this is implicit in the view that medicine is a vocation, though this description seems to have become less frequently used recently. Secondly, it should be noted that they place respect for the autonomy of the individual patient as the foremost of their basic principles. On reflection, the importance of such individual autonomy could be considered to have been previously neglected in engineering where the emphasis has been on benefits to society as a whole. This feature is related to a third, their incorporation of the ethics of care as a part of the overall analysis. This emphasis on the person will form a significant part of the aspirational engineering ethical ethos to be developed in the second part of the present book, though it will be traced to earlier sources than those used by Beauchamp and Childress.

A final element of the authors' analysis that is relevant to engineering occurs in their discussion of trustworthiness. Patients need to trust their doctors. Beauchamp and Childress observe that such trust is a fading ideal in contemporary health care institutions. Also, doctors are increasingly concerned as to whether they can trust their patients, as society becomes more litigious. Among the contributing causes of the loss of trust, particular attention is drawn to "the loss of intimate contact between physicians and patients, the increased use of specialists, and the large, impersonal, and bureaucratic medical institutions" (Beauchamp and Childress 2001: 35). As engineers frequently work at a distance in both place and time from those affected by their actions, as their roles may be highly specialised, and as they often work in large organisations, these are also likely to be relevant concerns for them. However, before moving explicitly to such considerations it is useful to consider aspects of the ethics of commercial enterprises, the ethics of business.

[3] In practice, doctors are often required to treat patients in accordance with strictly defined procedures.

4.3 Ethics in Business

There has been a concern for ethical practice in business from ancient times. For example, the Hebrew Bible includes the injunctions, "You shall not cheat in length weight or quantity. You shall have honest balances, honest weights, an honest ephah and an honest hin" (Leviticus 19.35)[4], and includes a specific ban on using different weights for buying and selling, "Differing weights are an abomination to the Lord" (Proverbs 20.23). It appears that regulation was needed even for the small-scale transactions carried out in such a predominantly agricultural society.

The influence of business on modern societies is immense. Large companies may have a level of economic activity comparable in value to that of a medium-sized nation state. The ethics of business has correspondingly become a very important issue. Further reasons for the strong focus on business ethics are that business malpractices have the potential to inflict enormous harm on individuals, on communities and on the environment, and that the demands being placed on business to be ethical by its various stakeholders are becoming more complex and more challenging (Crane and Matten 2004: 12).

In recognition of the influence of business on human wellbeing, the United Nations launched the *Global Compact* in 2000. The *Global Compact* is a network of commercial companies, six UN agencies, and labour and civil society organisations with thousands of participants from more than 100 countries. It does not have regulatory powers, relying instead on public accountability, transparency and the enlightened self-interest of the participants. For example, companies are expected to communicate publicly and prominently their actions taken toward Global Compact principles on an annual basis. These principles are a set of core values in the areas of human rights, labour standards, the environment and anti-corruption (UN 2008):

Human Rights

Principle 1 Businesses should support and respect the protection of international human rights within their sphere of influence; and

Principle 2 make sure they are not complicit in human rights abuses.

Labour

Principle 3 Businesses should uphold the freedom of association and the effective recognition of the right to collective bargaining;

Principle 4 the elimination of all forms of forced and compulsory labour;

Principle 5 the effective abolition of child labour; and

Principle 6 the elimination of discrimination in respect of employment and occupation.

---

[4] Scripture quotations are from the New Revised Standard Version of the Bible.

Environment

Principle 7	Businesses should support a precautionary approach to environmental challenges;
Principle 8	undertake initiatives to promote greater environmental responsibility; and
Principle 9	encourage the development and diffusion of environmentally friendly technologies.

Anti-Corruption

Principle 10	Businesses should work against corruption in all its forms, including extortion and bribery.

These principles aim to enjoy universal consensus and are based on the *Universal Declaration of Human Rights*, the *International Labour Organisation's Declaration of Fundamental Principles and Rights at Work*, the *Rio Declaration on Environment and Development* and the *United Nations Convention Against Corruption.* They indicate the great diversity of issues requiring consideration in developing a balanced view of business ethics. Aspects requiring consideration include both the economic and non-economic interests of the commercial enterprise, the employee, the customer, society as a whole and the environment.

Such a multifaceted approach to the effects of business is important if any initiative is to be really worthy of description as *ethical.* Not all approaches to business practices meet such a criterion. For example, a group of about 70 US companies building ships, aircraft, weapons and supporting systems are signatories of the *Defense Industry Initiative on Business Ethics and Conduct* (DII). These companies pledge to "adopt and implement a set of principles of business ethics and conduct that acknowledge and express their federal-procurement-related corporate responsibilities to the Department of Defense, as well as to the public, the Government and each other" (DII 2004). The initiative was developed as a response to fraud and allegations of fraud in this industry in the 1980s. It does not consider the nature of the industry's products. The DII is in fact essentially an agreement to follow federal procurement laws. Indeed, it might be considered to be reminiscent of an eighteenth century pirate's code, which included the article, "The captain and the quartermaster shall each receive two shares of a prize, the master gunner and boatswain, one and a half shares, all other officers one and one quarter, and private gentlemen of fortune one share each" (Breverton 2004). However, some actions may not be ethically virtuous however they are carried out.

In contrast, leading companies in other sectors set high standards for their staff in relation to all aspects of their activities. An example is provided by BP, a company with core activities primarily in the oil industry and employing about 100,000 staff in 100 countries. The company's Code of Conduct covers five key areas of their business operations (BP 2005):

health, safety, security and the environment – fundamental rules and guidance to help protect the natural environment, the safety of the communities in which they operate, and the health, safety and security of their people.

employees – covering fair treatment and equal opportunity, providing guidance for dealing with cases of harassment or abuse and for protecting privacy and employee confidentiality.
business partners – providing detailed guidance on giving and receiving gifts and entertainment, conflicts of interest, competition, trade restrictions, money laundering and working with suppliers.
governments and communities – covering such areas as bribery, dealing with governments, community engagement, external communications and political activity.
company assets and financial integrity – containing guidance about accurate and complete records and reporting, protecting company property, intellectual property, insider trading and digital systems.

The Code, which is detailed and available in eleven languages, prioritises the need for generating trust of customers, investors, employees, the communities in which they work and the societies in which they operate. There is a commitment to honesty, integrity and mutual advantage in all of the company's relationships. A company of the size of BP is subject to differing regulations throughout the world. The Code emphasises that where differences exist between the Code and local regulations or practices, whichever is the more stringent is to be applied.

BP employs many engineers, and for these the aspects of the Code dealing with health, safety, security and the environment will be of special relevance. The standard set is aspirational, stressing that no activity is so important that it cannot be done safely, with the goal of no accidents, no harm to people, and no damage to the environment[5]. BP has extensive procedures to support staff in the promotion of ethical conduct. Additionally, where an employee does not perceive these as providing a suitable source of help, a telephone and email facility operated by an independent company is available. This provides 24 hours a day/7 days a week advice on compliance and ethics, if necessary on a confidential basis. The text specifies that any employee who in good faith seeks advice, raises a concern or reports misconduct is regarded as following the Code. Correspondingly, failure to follow the Code is taken very seriously and may result in disciplinary action up to and including dismissal. Such action contrasts with the relatively weak regulation of conduct by professional engineering institutions (compared, for example, to the role of the GMC for doctors in the UK, as discussed earlier in this chapter). Three related salient aspects of the Code will be considered later – a clear statement that it does not serve as a substitute for individual responsibility for exercising good judgement (in addition to following the code), a commitment that each employee should feel supported in the management of their personal priorities, and a question suggested in cases of doubt about conduct, "What would others think about this action – your manager, colleagues or family?".

[5] Though even setting an aspirational standard can sometimes fail in practice, as shown by the serious explosion at a BP refinery in Texas in 2005, reportedly resulting in 15 people being killed and at least 170 people being injured.

4.4 Philosophical Business Ethics

Broadly, in business ethics there is a divide between compliance based approaches typical of the US and a European emphasis on aspirational approaches that are driven by values, also taking into account effects on individuals outside of the company and their communities. Therefore, in considering philosophical aspects of business ethics it is useful to look to work written from a European perspective, such as *Business Ethics* by Crane and Matten (2004). In assessing the applicability of traditional ethical theories to business ethics, the authors consider two consequentialist approaches, egoism and utilitarianism, and two non-consequentialist approaches, duty and rights/justice (Crane and Matten 2004: 79). The authors relate egoism to the market economics of the eighteenth century economist Adam Smith. It is an ancient philosophical approach, though some of the more recent traditional theories underrate the importance of self-interest. However, such an approach needs some regulating mechanism if it is to be effective. The authors note that markets cannot meet this role for two key aspects of modern business, globalization and sustainability. There is considerable doubt about the effectiveness of markets on a global level, and egoism takes no account of the effect of today's depletion of resources on future generations. In the case of utilitarianism they consider that it ultimately comes close to a cost-benefit analysis. They note that there is evidence that commercial managers continue to rely primarily on consequentialist thinking in making decisions. They consider the main problems in the application of utilitarianism to business to be the subjectivity of assessing consequences, the difficulty of assigning costs and benefits to every situation, and the overlooking of the interests of minorities. For example, it is difficult to apply utilitarianism in an unequivocal way to an assessment of the use of child labour.

In assessing non-consequentialist approaches, Crane and Matten (2001: 86) first consider Kant's maxims of duty and note that such a strong emphasis on human dignity would almost certainly rule out the use of child labour. The universalisation principle of the first maxim would rule out support for child labour in the third world unless it was also acceptable in Europe, which is not the case. Child labour also very likely violates the second maxim as it treats certain children as ends rather than respecting their integrity. More widely, the authors consider that application of Kant's approach to business has limitations due to its undervaluing of outcomes, the high demands it makes for abstract thinking and intellectual scrutiny, and its over-optimistic assessment of the role of rationality in decision making. Crane and Matten link rights and duties, as the rights of one person result in a corresponding duty for other persons to respect these rights. They trace a rights approach to the seventeenth century philosopher John Locke who identified rights to life, freedom and property. These and other rights feature strongly in modern thinking and legislation. Thus, Principle 5 of the *Global Compact* specified the abolition of child labour and Article 24 of the *Charter of Fundamental Rights of the European Union* states that in all actions relating to children the child's best interests must be a primary consideration. In the multicultural environment in which business operates there is, however, some concern that rights approaches are biased towards Western views of ethics. In considering justice, Crane and Matten focus on the approach of Rawls (1971/1999). Again using the example case of child labour they show how Rawls' analysis could be used either to rule out the use of child labour or to justify its use in the third world by a multinational company, depending on how the principle of fairness is applied. An even more basic problem in the application of contractual analyses such as Rawls' to a case like child labour is that the most disadvantaged, the children, are excluded from the formulation of any agreement.

Among their criticisms of ethical theories when applied to business are that they are too abstract, too reductionist, too objective and elitist, too impersonal, and too rational and codified. These factors lead Crane and Matten to consider contemporary (or revived) theories that are more flexible and give greater emphasis to the role of individuals and to the decision making context. As important approaches that appear less commonly in business ethics texts they propose virtue ethics, feminist ethics, discourse ethics and postmodern ethics (Crane and Matten 2004: 96). They note the revival of Aristotelian virtue ethics, especially as proposed by MacIntyre (1981/1985) with the community in which the virtues are learnt being the business community rather than the city-state. Such an approach is less prescriptive than those usually applied to business, but it has the advantage through its emphasis on the cultivation of good judgement of providing a means of assessing situations that are not amenable to specific rules or principles. Feminist ethics has drawn attention to the significance of care in ethics. (Its more considered forms have recognised that such an approach is not intrinsically connected with being female (Baier 1994/1997: 277). The modern recognition of the immense importance of care in ethics could also be considered to be a revival of the ethics found in many religious traditions, and in some parts of the world it might be more productive to present it in this way.) Discourse ethics is an approach to reaching ethical agreement that in business would emphasise the importance of reaching a consensus view in a peaceful way. It is more a way of doing ethics rather than an ethical theory and certainly has value in multicultural environments where agreement needs to be reached across conceptual boundaries. Postmodern ethics (Bauman 1993) has similarities with virtue ethics. It emphasises the responsibility of the individual rather than the role of overriding principles. In difficult business situations it would have the advantage of promoting modesty in all concerned about the priority of their convictions over those of others. Aspects of virtue ethics, the ethics of care and of postmodern ethics will play roles in the aspirational approach to engineering ethics to be developed in the second part of the present book.

In considering the relevance of business ethics to such a development of engineering ethics it is informative to consider work on cognitive moral development (CMD) to which Crane and Matten draw attention. CMD provides an analysis of the level at which reasoning about ethical issues is carried out. The most widely discussed analysis identifies three levels divided into six stages (Kohlberg 1969: 376):

Level 1
Stage 1 Obedience and punishment orientation
Egocentric deference to superior power or prestige, or a trouble-avoiding set. Objective responsibility.
Stage 2 Naively egoistic orientation
Right action is that instrumentally satisfying the self's needs and occasionally others. Awareness of relativism of value to individual needs and perspective. Naïve egalitarianism and orientation to exchange and reciprocity.

Level 2
Stage 3 Good-boy orientation
Orientation to approval and to pleasing and helping others. Conformity to stereotypical images of majority or natural role behaviour, and judgement by intentions.
Stage 4 Authority and social-order maintaining orientation
Orientation to "doing duty" and to showing respect for authority and maintaining the given social order for its own sake. Regard to earned expectations of others.

Level 3
Stage 5 Contractual legalistic orientation
Recognition of an arbitrary element or starting point in rules or expectations for the sake of agreement. Duty defined in terms of contract, general avoidance of violation of the will or rights of others, and the majority will and welfare.
Stage 6 Conscience or principle orientation
Orientation not only to actually ordained social rules but to principles of choice involving appeal to logical universality and consistency. Orientation to conscience as a directing agent and to mutual respect and trust.

Initial empirical investigation of managers indicated that they typically carried out moral reasoning at stages 3 or 4 (Weber 1990). However, managers employed in large or medium-sized organisations tended to reason at lower stages than managers who worked in small firms or were self-employed. Moral reasoning exhibited when the dilemmas were placed in a business context was at a significantly lower level than for a dilemma in a non-business context. It was thought that feelings such as "I'm just a cog in the machine" or "I just don't know all the facts and I will trust that my superior knows what he is doing" were more common in larger organisations. Such managers may have perceived their "world" as very small, limited to a division or department. Managers in small organisations were thought to have different perception and behaviour as they were more aware of the larger "world", including customers and the public. However, later studies have shown considerable complexity and that most moral reasoning research is inflated due to the asking of socially desirable or hypothetical questions (Weber and Gillespie 1998; Weber and Wasieleski 2001)[6]. When subjects are asked what stages they "really use" the responses are more often at stages 2 and 3. Context is very important, and an

[6] I thank Professor Weber for helpful advice.

individual may operate at any of the stages depending on the situation. For example, self-preservation may be dominant in a directly life-threatening situation. There is also a variation with industry type, with health care managers having the highest stage scores and construction industry managers the lowest.

These findings may well be evidence that the more abstract traditional ethical theories, which tend to lie at stages 5 and 6, are for many too complex, are unable to provide motivation, or appear impractical. For this reason, the modern revitalisation of virtue approaches may be seen as being especially promising. Indeed, a values based culture shift approach has been recognised as being very effective at raising ethical awareness in companies. Any development of an aspirational engineering ethical ethos needs to be formulated in an appropriate way if it is to be capable of being used practically.

Crane and Matten opened their analysis of business ethics by noting that everyday business activities require the maintenance of basic ethical standards, such as honesty, trustworthiness and cooperation. However, they also note that people often "bracket" their personal ethical values while they are at work. This susceptibility to suppression of ethical responsibility seems to be a feature of goal- and end- oriented work environments. People are employed for their specialised technical know-how with a related tendency to convert issues of ethical disagreement into issues of technical disagreement. In simple terms, they have only a role.

Bauman has identified two ways in which organisations suppress individual ethical responsibility (Bauman 1994: 7). The first, "floating responsibility", concerns the difficulty of specifying where responsibility lies. In effect, each employee can be considered to be performing actions that in isolation might arguably be innocuous and it is only through the complementary actions of many others that deleterious outcomes are produced. The second, "moral neutralisation", is the tendency to declare that most activities of employees are exempt from moral evaluation, the only valid standard being procedural correctness. In response, postmodern analyses have stressed the need for a return to an holistic approach emphasising the need for coherence in an individual's personal and professional activities in recognition of their intimate connections (Bauman 1994: 35). In this respect, the three salient features of the BP Code of Conduct mentioned earlier, which identified personal responsibility, a commitment to support of an employee's personal priorities, and the key question relating professional decisions to ethics in the context of family, represent a very enlightened approach to business ethics. These are certainly features that warrant inclusion in an aspirational engineering ethic. They represent an important counterbalance to the spatial and temporal distance that can separate an individual's actions from their final effects in both business and engineering.

References

Baier A (1997) What do women want in a moral theory?. In: Virtue ethics, Crisp R and Slote M (eds), Oxford University Press, Oxford, p. 262–277. Originally published 1994
Bauman Z (1993) Postmodern ethics, Blackwell, Oxford
Bauman Z (1994) Alone again: ethics after uncertainty, Demos, London
Beauchamp TL, Childress JF (2001) Principles of biomedical ethics, 5th edition, Oxford University Press, Oxford
Breverton T (2004) Black Bart Roberts, Glyndŵr Publishing, St Athans
British Medical Association (2001) Recommendations for the medical profession and human rights: handbook for a changing agenda, BMA, London
British Medical Association (2007a) The use of drugs as weapons, BMA, London
British Medical Association (2007b) Medical milestones: celebrating key advances since 1840, British Medical Journal 334. Poll results are given at http://www.bma.org.uk/ap.nsf/Content/BMJPollresults
BP (2005) BP code of conduct, BP, London
Crane A, Matten D (2004) Business ethics, Oxford University Press, Oxford
Defense Industry Initiative (2004) The defense industry initiative on business ethics and conduct, DII, Washington
General Medical Council (2006) Good medical practice, GMC, London
Kohlberg L (1969) Stage and sequence: the cognitive-developmental approach to socialisation, in Handbook of socialisation theory and research, Goslin DA (ed), Rand McNally and Company, Chicago, p. 347–480
MacIntyre A (1985) After virtue, 2nd edition, Duckworth, London. First published 1981
Rawls J (1999) A theory of justice, revised edition, Oxford University Press, Oxford. First published 1971
United Nations (2008) http://www.unglobalcompact.org/AboutTheGC/TheTenPrinciples/index.html
Weber J (1990) Managers' moral reasoning: assessing their responses to three moral dilemmas, Human Relations 7: 687–702
Weber J, Gillespie J (1998) Differences in ethical beliefs, intentions and behaviours: the role of beliefs and intentions in ethics research revisited, Business and Society 37: 447–467
Weber J, Wasieleski D (2001) Investigating influences on managers' moral reasoning, Business and Society 40: 79–111
World Medical Association (2006a) Declaration of Geneva, WMA, Ferney-Voltaire
World Medical Association (2006b) Statement on weapons of warfare and their relation to life and health, WMA, Ferney-Voltaire

Chapter 5
Reflection

Accounts of both engineering and ethics tend to be written in rather formal technical prose. A major advantage of such prose is that it facilitates great conceptual precision in communication between specialists familiar with the terminology used. A serious disadvantage of such prose is that as it develops in sophistication it can become increasingly less concrete. In particular, the terminology can disguise the real human significance of the issues discussed.

Many means of communication can express matters of human significance more effectively than technical prose. Music and art are powerful examples. There are also many forms of the written word which have greater expressive range. The first aim of the present chapter is to provide brief expressions of some issues related to the use of engineering using forms of the written word that are more evocative than formal technical prose. The second aim is to provide corresponding brief expressions of attitudes related to the aspirational engineering ethos to be presented in the subsequent chapters.

5.1 On War

The poet Wilfred Owen was killed in military action on 4 November 1918. Earlier that year he had drafted the Preface for a collection of poems that he had hoped to publish in 1919 (Owen 1920/1967):

> This book is not about heroes. English poetry is not yet fit to speak of them.
> Nor is it about deeds, or lands, nor anything about glory, honour, might, majesty, dominion, or power, except War.
> Above all I am not concerned with Poetry.
> My subject is War, and the pity of War.
> The Poetry is in the pity.
> Yet these elegies are to this generation in no sense consolatory. They may be to the next.
> All a poet can do today is warn. That is why the true Poets must be truthful.

Regrettably, the warnings of the poets of the First World War have not been heeded. Preparation for war takes place on an ever-increasing scale, facilitated by the work of engineers in the design, manufacture and use of weapons of huge devastation potential. These engineers do not appear to properly visualise the effects of their work. It seems that they are lost in the labyrinth of technology, dazzled by technical innovation.

One of Owen's most moving poems, *Disabled*, is about the devastation of an individual who had been seduced by a dazzling presentation of the "thrill" of military engagement. It is a poem that merits reflection by all engineers working on military projects, for it is a poignant account of the fate that they may be preparing for others (Owen 1920/1967):

He sat in a wheeled chair, waiting for dark,
And shivered in his ghastly suit of grey,
Legless, sewn short at elbow. Through the park
Voices of boys rang saddening like a hymn,
Voices of play and pleasure after day,
Till gathering sleep had mothered them from him.

About this time Town used to swing so gay
When glow-lamps budded in the light blue trees,
And girls glanced lovelier as the air grew dim, –
In the old times, before he threw away his knees.
Now he will never feel again how slim
Girls' waists are, or how warm their subtle hands;
All of them touch him like some queer disease.

There was an artist silly for his face,
For it was younger than his youth, last year.
Now, he is old; his back will never brace;
He's lost his colour very far from here,
Poured it down shell-holes till the veins ran dry,
And half his lifetime lapsed in the hot race
And leap of purple spurted from his thigh.

One time he liked a blood-smear down his leg,
After the matches, carried shoulder-high.
It was after football, when he'd drunk a peg,
He thought he'd better join. – He wonders why.
Someone had said he'd look a god in kilts,
That's why; and may be, too, to please his Meg;
Aye, that was it, to please the giddy jilts
He asked to join. He didn't have to beg;
Smiling they wrote his lie; aged nineteen years.
Germans he scarcely thought of; all their guilt,
And Austria's, did not move him. And no fears
Of Fear came yet. He thought of jewelled hilts
For daggers in plaid socks; of smart salutes;
And care of arms; and leave; and pay arrears;
Esprit de corps; and hints for young recruits.
And soon, he was drafted out with drums and cheers.

> Some cheered him home, but not as crowds cheer Goal.
> Only a solemn man who brought him fruits
> *Thanked* him; and then inquired about his soul.
>
> Now, he will spend a few sick years in institutes,
> And do what things the rules consider wise,
> And take whatever pity they may dole.
> To-night he noticed how the women's eyes
> Passed from him to the strong men that were whole.
> How cold and late it is! Why don't they come
> And put him into bed? Why don't they come?

It is not only those who are killed who suffer in war. Though this young man has suffered horrific physical injury, it his mental torment that is emphasized rather than his pain. Owen sympathises strongly with the young man who has been seduced by the glorification of war in recruitment propaganda. The glorification of war is still used to recruit soldiers. The glorification of the technical achievement of advanced weaponry is used to recruit engineers to military projects.

Modern engineers are, of course, much less directly involved in devastation than was the subject of Owen's poem. They often engage in military work with virtually zero risk to their own wellbeing. Indeed, this lack of proximity is an important factor in their continued activities – they are essentially completely isolated from the results of their work. They can "bracket" such results from their everyday activities. They do not see the casualties of present warfare, including the 90% who are (civilian) women and children.

In the years preceding the Second World War, the callousness that lack of proximity can engender was recognised by the poet Dylan Thomas, an admirer of Owen's verse. In a newspaper article, Thomas wrote about the imbalance of society in the 1930s, "the society that feeds one member and starves another, that increases its murderous air-fleet rather than its dole, is a society with its roots in filth and injustice". He also wrote of the fundamental misrepresentation of human nature and the false representation of the nature of war by, "the militarist who explains to a school of boys that man is naturally a fighting animal, that war is a glorious adventure, and that the great men of history are always butchers" (Thomas 1934). Thomas further expressed his views in a short poem, *The Hand that Signed the Paper* (Thomas 1933/1952), which describes the human and material devastation that may be caused by the signing of a political directive to begin a war or by the signing of a commercial armaments deal. By such means, even those who are far from the scene of violent actions can unleash great devastation. The poem ends with the memorably evocative words, "Hands have no tears to flow". However, it is not only politicians or business leaders who can be responsible in this way. In technological times, engineers have a great part to play in the remote causation of destruction. Lack of proximity engenders lack of compassion. If hands have no tears to flow, then nor does advanced technology such as computer based weapons simulations or laser-guided bombs.

5.2 On Hunger and Thirst

Engineering has a major role to play in helping to provide safe food and water. Contemporary experiences of extreme hunger are fortunately rare in Western Europe. However, events such as the Great Irish Famine in the middle of the nineteenth century still feature in the collective consciousness. It is estimated that about 1 million people died and 1 million people emigrated from Ireland as a result of this famine, in total about a quarter of the population. Most people in Ireland were very poor at the time and relied heavily on potatoes for food, and crop blight was the immediate trigger of the famine. Seamus Heaney recalls the event in the third section of a long poem, *At a Potato Digging* (Heaney 1966), which includes the words, "Mouths tightened in, eyes died hard, faces chilled to a plucked bird ... A people hungering from birth, grubbing like plants, in the bitch earth, were grafted with a great sorrow. Hope rotted like a marrow." The consequences of the famine seem to have been greatly exacerbated by the actions, or lack of actions, of the English authorities – means were available that could have alleviated the burden of human suffering. Indeed, it seems that English landowners were exporting from Ireland wheat and other staple crops whilst the indigenous population starved. We stand in an analogous position today regarding the lack of water for the cultivation of food, and the lack of drinking water and sanitation, experienced by many of the world's population. We have the means available to alleviate the suffering but do not apply them.

Extreme thirst has been seldom experienced in Europe during the course of everyday life. However, Samuel Taylor Coleridge expresses some of the suffering and delirium experienced due to thirst in verses from *The Rime of the Ancient Mariner* (Coleridge 1798/1972):

> Water, water, every where,
> And all the boards did shrink;
> Water, water, every where,
> Nor any drop to drink.
> ...
> And every tongue, through utter drought,
> Was withered at the root;
> We could not speak, no more than if
> We had been choked with soot.
> ...
> There passed a weary time. Each throat
> Was parched, and glazed each eye.
> A weary time! a weary time!
> How glazed each weary eye,
> When looking westward, I beheld
> A something in the sky.
> ...

With throats unslaked, with black lips baked,
We could nor laugh nor wail;
Through utter drought all dumb we stood!
I bit my arm, I sucked the blood,
And cried, A sail! a sail!

Though in extreme thirst, the Ancient Mariner was surrounded by plenty of water, the sea. Likewise, much of the world's population lives within a few kilometres of the sea. Modern processes such as membrane desalination have a vital role to play in providing such populations with drinking water. Indeed, a Modern Mariner would have had such an engineered process available on his ship.

5.3 Responsibility and Aspiration

Important themes of the present book are the need for engineers to take greater personal responsibility for their actions and the need for engineers to acquire an aspirational ethos. These themes have been considered in literature in many ways. They will be illustrated here with two texts, one from the Hebrew Bible and one from the New Testament. To some readers it may appear curious to include texts that are religious in origin in a work about engineering. However, these examples are laconic classics, richly expressing universal concerns in very few words.

The first is the story of David and Bathsheba. King David was an iconic figure in the history of ancient Israel. This part of his narrative starts unpretentiously:

> It happened, late one afternoon, when David rose from his couch and was walking on the roof of the king's house, that he saw from the roof a woman bathing; the woman was very beautiful. (2 Samuel 11.2)[1]

This simple beginning leads to events that have serious repercussions throughout the rest of David's life. David becomes infatuated with the woman, Bathsheba. He uses his role as king to ensure that her husband, Uriah, is killed in battle – killed by agents at a distance. David can then make her his wife.

The story continues with the prophet Nathan being sent to David. Nathan tells him a parable about a rich man, who has many sheep and cattle, and a poor man, who has just one sheep. The rich man takes this single sheep and slaughters it when he receives an unexpected visitor. David responds angrily:

> As the Lord lives, the man who has done this deserves to die; he shall restore the lamb fourfold, because he did this thing, and because he had no pity. (2 Samuel 12.5–6)

[1] Scripture quotations are from the New Revised Standard Version of the Bible.

And the response from Nathan is direct:

> Nathan said to David, "You are the man!". (2 Samuel 12.7)

The brief outline omits many subtleties, for such Hebrew stories seldom make simple moral points like some Greek fables. Nathan's parable, for example, does not exactly match David's actions. However, even a simplified version of the story has several features that it would be well to bear in mind in developing an ethic of engineering:

- David's murderous course of action starts with an everyday event, a stroll in the evening. There is no initial ethical quandary.
- An ancient king was in a position of great influence and in particular had an obligation to promote justice, especially to the disadvantaged. David fails to meet these responsibilities.
- Despite his official role, David is personally responsible, even though Uriah was killed at a distance and by enemy soldiers, and he recognises this himself in his later response to Nathan's indictment.

As engineers we are often responsible *to* others. We should also consider ourselves responsible *for* others. In ancient times, this responsibility for others was expressed as love for one's neighbour, for others in one's group. The New Testament gives an account of a lawyer seeking a clearer specification of this responsibility by asking, "Who is my neighbour?", very likely with the aim of trying to limit his liability. Jesus replies by telling the following story, which has become known as the parable of the Good Samaritan:

> "A man was going down from Jerusalem to Jericho, and fell into the hands of robbers, who stripped him, beat him, and went away, leaving him half dead.
> Now by chance a priest was going down that road;
> and when he saw him, he passed by on the other side.
> So likewise a Levite,
> when he came to the place and saw him, passed by on the other side.
> But a Samaritan while travelling came near him;
> and when he saw him, he was moved with compassion.
> He went to him and bandaged his wounds, having poured oil and wine on them. Then he put him on his own animal, brought him to an inn, and took care of him. The next day he took out two denarii, gave them to the innkeeper, and said, 'Take care of him; and when I come back, I will repay you whatever more you spend.'
> [Then Jesus asked] Which of these three, do you think, was a neighbour to the man who fell into the hands of the robbers?" He [the lawyer] said, "The one who showed him mercy." Jesus said to him, "Go and do likewise." (Luke 10.30–37)

It is a feature of writing such as this that the realisation of meaning is not merely *illustrated* in the story form but *constituted* by it. However, it can be helpful for the present purposes to draw attention to some significant features:

- The robbers strip the traveller, so that neither his national identity nor his social status is apparent. He is simply a person in need.
- The priest and the Levite hold positions of authority and responsibility. They are subject to religious codes of action, which theological analysis shows to have contradictory requirements in this case: whether to show love of neighbour or avoid ritual uncleanness through contact with a (possibly) dead body.
- Only the Samaritan, an outsider with no special responsibility, recognises the need of the injured man. His compassion moves him to action – he does not carry out a consequentialist calculation, act because of his contractual or legal obligation, or act purely from a sense of duty.
- The Samaritan is not just minimally helpful. He acts in a way that ensures the injured man's recovery.
- At the end of the story, the neighbour is redefined as the one who acts, not as the one who has needs.

The parable has been described as "a paradigm of the compassionate vision which is the presupposition for ethical action" (Donahue 1988: 132). Most importantly, it constitutes an account of a universal ethic of responsibility for any person in need.

References

Coleridge ST (1972) The rime of the ancient mariner. In: The new Oxford book of English verse, Gardner H (ed), Oxford University Press, Oxford. Poem originally published 1798

Donahue JR (1988) The gospel in parable, Fortress Press, Philadelphia

Heaney S (1966) Death of a naturalist, Faber, London

Owen W (1967) The collected poems of Wilfred Owen, Lewis CD (ed.), Chatto and Windus, London. Poems first published 1920

Thomas D (1934) A plea for intellectual revolution, Swansea and West Wales Guardian, 3 August 1934 (both quotes). Cited in Maud R (2003) Where have all the old words got me?, University of Wales Press, Cardiff, p. 238

Thomas D (1952) Collected poems 1934–1952, Dent JM and Sons, London. Notebook version August 1933

Part II

Chapter 6
Outline of an Aspirational Engineering Ethics

6.1 Introduction

Engineering has made an enormous contribution to providing a basis for human flourishing. The provision of benefits such as clean water, medicines, safe foods, energy and transport gives rise to a basic level of wellbeing that frees us to enjoy the personal, cultural and intellectual activities that can characterise a fulfilled life. However, engineers are prone to become lost in the labyrinth of technology, to prioritise technical ingenuity over helping people. As a consequence, there are tendencies in a number of cases to develop and apply complex but inappropriate technology, whereas at the same time technology that could alleviate great human suffering is not being applied as extensively or as rapidly as it could be. Such grave imbalances in the use of skills and resources should be a cause of concern to all engineers.

Engineering ethics is a relatively underdeveloped subject and has only recently been given broad explicit consideration within the engineering profession. However, as engineering is a practical subject that has a great impact on human lives, engineers have always had to make ethical decisions, though the ethical viewpoints that have underlaid such decisions have generally remained implicit. Analysis of these implicit ethical viewpoints shows that they have been most often based on what in philosophical terms would be characterised as consequentialism, contractualism and duty, with some influence also of ideas of virtue. Engineers who have written explicitly about professional ethics have also mostly formulated their viewpoints in these terms.

Explicit consideration of professional ethics has a much longer tradition in medicine and has also received great attention in business. Approaches to ethics in both medicine and business have much to offer the development of engineering ethics. The wellbeing of the individual patient has been paramount in medicine and the priority of personal care is promoted by the close proximity, in both place and time, of doctor and patient. However, there has recently been a tendency to loss of trust related to an increase in size of medical organisations and a decrease in intimate contact between doctors and patients. The maintenance of high ethical standards has been a major concern of business ethics and some leading international companies have adopted active, aspirational approaches. Two findings of sociologists warrant special attention in this case: the tendency for individuals to bracket their personal ethical values whilst at work and the tendency of employees to restrict their ethical reasoning to a level of conformity to acceptable expectations rather than to aspire to the highest achievable levels.

The formulation of an aspirational engineering ethical ethos needs to take into consideration the current imbalance in engineering priorities, the limitations in the nature of the philosophical analyses previously applied to engineering ethics and the lessons arising from the more advanced states of medical and business ethics. There is a particular need to restore the priority in engineering of helping people over technical ingenuity for its own sake. This implies the need to develop a means of promoting a sense of proximity between the engineer and those affected by engineering activities. There is also a need to find a formulation and vocabulary that are less formidable than those of the traditional mainstream philosophical writings. Ideally, such a formulation would build on the personal commitment to ethical behaviour that most people exhibit in their personal lives. In this context it is worth noting that a number of philosophers have also been concerned that the mainstream of philosophical ethics has been too abstract, "What is missing in these theories is simply – or not so simply – the person" (Stocker 1997: 71), and that their emphasis on systems over people can lead to serious injustices (Anscombe 1958/1997: 44).

The present chapter will develop an outline of an aspirational engineering ethical ethos in three stages. The first stage will emphasise the priority of people in engineering ethics in a way that also gives due attention to the necessity of technical achievement in professional activities. This stage will use Buber's seminal vocabulary for describing our existence using the "primary words" *I-Thou* and *I-It*. Consideration will also be given to Levinas' more radical description of the ethical priority of others in determining our actions, a notion he terms *the face*. The second stage will consider ways of promoting continuity and coherence in engineering ethics. In the context of the engineering profession, this will be approached by using MacIntyre's concept of a *practice*. For an individual, the coherence of personal and professional ethics will be approached through the concept of the *narrative unity* of a life. The third stage will consider the nature of and relationship between three types of care, *I* (self care), *I-Thou* (proximate care) and *I-You* (care beyond proximity). Achieving a balance of such commitments is a challenge for all in the modern world, and especially so for professionals such as engineers.

6.2 The Priority of People

In the history of philosophical ethics there have been pessimists, who viewed the nature of people as being essentially selfish and ethics as being something which needed to be imposed to allow the possibility of mutual survival, and optimists, who emphasised the social nature of people and their natural tendency to thrive through compassionate interaction. Recent studies of the evolution of ethics confirm the social nature of humans and that helpful behaviour is an essential part of our constitution. The presence of violence in human societies shows that there is also a darker side to our nature. The challenge for an aspirational ethical ethos is to find ways to express and promote the sympathetic side of our nature. In the context of engineering ethics it is beneficial if such expression also provides a means of describing its relationship to our technical activities.

A means toward such expression has been provided by Buber's description of human activities in terms of the "primary words" *I-It* and *I-Thou* (Buber 1923/2004). Buber was concerned with an individual's relationship to other people and to the natural world. It is especially useful in the present context that he develops a philosophical approach to these relationships that is based on everyday experience of dealing with others and the physical world.

The world of *I-It* is the world of *experiencing* and *using* with the aim of "sustaining, relieving and equipping of human life" (Buber 1923/2004: 36). This includes the world in which the engineer would usually be seen as carrying out his or her professional tasks, the world of mathematical analysis and physical exploitation of materials and processes. The world of *I-It* is where a man or woman:

> ... works, negotiates, bears influence, undertakes, concurs, organises, conducts business, officiates ... the tolerably well-ordered and to some extent harmonious structure, in which, with the manifold help of men's brains and hands, the process of affairs is fulfilled. (Buber 1923/2004: 39)

Buber observes that both in the life of an individual and with the progress of history there is a progressive augmentation of the world of *It*, "the world of objects in every culture is more extensive than that in its predecessor" (Buber 1923/2004: 35). Indeed, the world of knowledge is structured around *It*, and this is essential to our existence: "You cannot hold on to life without it, its reliability sustains you" (Buber 1923/2004: 31). However, in Buber's analysis the world of *I-It* is severely limited, "he who lives in *It* alone is not a man" (Buber 1923/2004: 32) and indeed holds inherent dangers: "If a man lets it have the mastery, the continually growing world of *It* overruns him and robs him of the reality of his own I" (Buber 1923/2004: 41) for, "The primary word *I-It* can never be spoken with the whole being" (Buber 1923/2004: 11). That is, fulfilment in life, living a fully human life, comprises more than just experiencing and using.

The world of *I-Thou*[1] is characterised by *meeting*. As will already be apparent, Buber's writing is more poetic than is usual in philosophical works. Hence, he illustrates his core meaning in the following progression:

> Consider a tree ... I can look on it as a picture ... I can perceive it as movement ... I can classify it in a species and study it as a type ... I can subdue its actual presence and form so sternly that I recognise it only as an expression of law ... I can dissipate it and perpetuate it in a number ... In all this the tree remains my object ... It can, however, also come about, if I have both will and grace, that in considering the tree I become bound up in relation to it. The tree is no longer an *It*, I have been seized by the power of its exclusiveness. (Buber 1923/2004: 14)

What is significant here is the possibility of a close relationship engendering *care* for the other. It should be noted that Buber chooses a tree, an inanimate object, as his primary example. In the case of a person the relationship is more intimate, for the response of the other may be explicit.

Further, Buber notes that, "The relation to the *Thou* is direct. No system of ideas, no foreknowledge, and no fancy intervene between *I* and *Thou*" (Buber 1923/2004: 17). Most importantly, "The primary word *I-Thou* can be spoken only with the whole being" (Buber 1923/2004: 17). This is the essence of a fully human way of being. However, vigilance is always necessary and especially so in highly technological societies, *for the development of experiencing and using comes about mostly through the decrease of an individual's power to enter into relation*. Even when relation has been fulfilled there remains the danger that the *Thou* will degenerate to an *It*.

Buber's unconventional writing style makes demands for imaginative reflection on the part of the reader. However, it provides a number of insights of value in the development of an aspirational engineering ethical ethos. It is clear that Buber has provided a vocabulary and approach that enables a discussion of important aspects of human life. The *I-It* world would be recognised by engineers as a description of the conditions under which they commonly carry out their professional activities. Buber rightly emphasises the importance of technical knowledge and technical activities. The development of the *I-It* concept also provides an account of the unease that they may feel about being restricted to technical aspects alone in the course of their employment. *Buber's description of the progressive augmentation of the world of It is a means of expressing the danger of becoming lost in the labyrinth of technology.*

Buber's description of *I-Thou* interactions has the exceptional advantage of encompassing both person/person and person/natural world (environmental) relationships.

---

[1] "Thou" corresponds to "Du" in the original German or to "Tu" in French. Its use has fallen out of modern English. However, just as the corresponding words in modern German and French, it was earlier used to indicate a close relationship with the person addressed.

Engineers would certainly recognise *I-Thou* experiences in their interpersonal relationships and most likely in their interaction with the natural world. However, they may not have considered such *I-Thou* experiences as being part of their engineering activities. This vocabulary and approach should encourage engineers to re-evaluate their attitudes to the ethics of their professional activities. This could occur through the *I-Thou* relation providing a conceptual grounding for desirable goals. For example, the *I-Thou* concept can express concern for the individual persons affected by such activities, rather than the statistical concern of consequentialism and contractualism or the impersonal concern of duty. Further, in dealing with the natural world there is a great awareness among engineers of the environmental consequences of their activities, but environmentalism may appear an unjustified fundamentalism if not supported conceptually in such a way.

Most importantly, Buber's approach provides a vocabulary that can act as a basis for an ethos promoting high ethical aspirations. An ethos expressed in this way has the potential for being recognised and accepted across the boundaries of cultures and beyond the minimum requirements of national or international law. The vocabulary and approach provide means for engineers to express in dialogue and cooperation a new prioritisation for the application of their skills.

In order to emphasise the importance of individual people in ethical analysis it is worth considering an observation on the viewpoint of moral scepticism. A much discussed account of this viewpoint is Mackie's *Ethics: Inventing Right and Wrong* which opens with the words, "There are no objective values" (Mackie 1977: 15). A key feature of Mackie's argument for moral scepticism is his "argument from queerness"; "If there were objective values, then they would be entities or qualities or relations of a very strange sort, utterly different from anything else in the universe" (Mackie 1977: 38). This argument is based on an emphatically physical sciences view of the world. On the basis of a sophisticated Kantian analysis, Korsgaard has made a highly perceptive response to this scepticism by taking up the use of the term "entities": "For it is the most familiar fact of human life that the world contains entities that can tell us what to do and make us do it. They are people, and the other animals" (Korsgaard 1996: 166).

Korsgaard's statement of the demands of others, who can tell us what to do, is more forcefully expressed than Buber's formulation. An even stronger formulation has been provided by Levinas. He notes that we do not live as isolated beings for whom the events in the world are a spectacle but that we are engaged in the world, and our primary engagement is ethical, that is, it involves relation to other people. Levinas reverses the philosophical tradition that has seen ethics as arising initially out of self-interest by instead prioritising the demands that an other can make on us, which he designates by the notion of *the face*. He describes an ethical act as being defined by "a response to the being who in a face speaks to the subject and tolerates only a personal response" (Levinas 1961/1969: 219). This view is based on the proximity of others generating a sensibility that leads to their needs taking priority. He even writes of the needs of others making us hostages, that we are above all responsible for others, indeed summoned to responsibility, even to the degree of substituting their needs for ours (Levinas 1974/1981: 117, 184). Levinas does not develop his philosophical arguments in a conventional way but his message is nevertheless clear. In simple terms, and changing the metaphor, we need to hear the voice of others saying, "It's me here, please help me!". According to Levinas the individual's response should be, "here I am for the others" (Levinas 1974/1981: 185).

Both Buber and Levinas recognise the essentially social nature of humans; concern for others is not something that needs to be added to an individual's existence by means of a contract or sophisticated analysis of the consequences of actions. Levinas' notion of responding to *the face* of others provides a strong visualisation of the nature of personal responsibility. It certainly applies to those we encounter closely and is further reflected in the strong response we have even to photographic images of others suffering. However, his approach is idealistic rather that aspirational – only a saint would be likely to be motivated solely by the needs of others.

It may also be considered that Buber's notion of *I-Thou* relationships demands too much. It assumes a close proximity and there is certainly a limit to the number of such relationships that an individual can maintain. Hence, it has been proposed that a third type of interaction, an *I-You* relationship, may be an appropriate description of many encounters in modern society (Cox 1965: 61). Such relationships would fully acknowledge the significance of the individuals with whom we interact but also recognise that the interaction lacks intimacy and is transitory. Such *I-You* relationships are a common feature of modern urban life. An *I-You* vocabulary can also provide an expression of the distinction between proximate care and care beyond proximity that is a necessary feature of engineering ethics and which will be considered later in the present chapter.

6.3 Continuity and Coherence: the *Practice* of Engineering

There has been a tendency for accounts of ethics to focus on ethical dilemmas, leading to what could be termed a "quandary" view, or even a "catastrophe" view, of ethics. Such a tendency has certainly been dominant in published accounts of engineering ethics. Alternatively, it might be proposed that the key ethical features of an individual's life are not responses to quandaries but rather the adoption of a positive way of life the very nature of which helps to minimise the occurrence of such difficult situations. Examples of such an approach have already been considered in the account of Aristotle's approach to ethics and of Franklin's principles of virtue given in Chap. 3.

A virtue approach to ethics requires careful consideration in the development of an aspirational engineering ethical ethos. A focus on virtues is certainly a way of expressing and promoting the sympathetic side of our nature using a vocabulary that is readily understood. However, there is one feature of virtue ethics that requires special attention if it is to be successfully applied: that the various proponents of this approach provide different listings and prioritisations of qualities considered to be virtues. MacIntyre (1981/1985) has traced the historical development of ideas of virtues. Like Buber and Levinas, he has argued that traditional approaches have placed too little emphasis on persons, and further that such approaches also placed too little emphasis on the contexts of their lives. MacIntyre noted that in earliest times, such as described in Homer's *Iliad* and *Odyssey*, the virtues most valued were physical strength, courage, cunning and friendship. As already noted, in a somewhat later period of Greek culture, Aristotle rated justice and friendship as the greatest virtues, but also valued courage, temperance, liberality, magnificence, pride, honour, good temper and tact. Closer to modern times, Franklin valued temperance, silence, order, resolution, frugality, industry, sincerity, justice, moderation, cleanliness, tranquillity, chastity and humility. MacIntyre sees these differences as reflecting the social circumstances in which they arose. Thus, Homer lived in a society that was vulnerable and very aware of the fragility of human life. Aristotle lived in a city-state in which the relationship between being a good man and being a good citizen was central. According to MacIntyre, Franklin's cultivation of virtues was intended to promote "happiness understood as success, prosperity in Philadelphia and ultimately in heaven" (1981/1985: 185).

Mention of heaven brings to mind the influence of religion on concepts of virtues, an influence that has been given rather too little credit in the modern resurgence of interest in virtue approaches to ethics. Thus, in the Christian tradition, Paul writes of the virtues of faith, hope, and especially love (1 Corinthians 13.13). He also writes of love, joy, peace, patience, kindness, generosity, faithfulness, gentleness and self-control (Galatians 5.22). The formation of communities based on mutually supportive fellowship was a key aspect of Paul's work. Buddhism also lays great importance on the cultivation of virtuous dispositions with the three key virtues identified as detachment, benevolence and understanding. Other important Buddhist virtues include compassion, generosity, and *ahimsã*, a "deeply positive feeling of respect for all living things" (Keown 2005: 12–16). Confucius gave great priority to the cultivation of virtues. Thus, when asked how a man of integrity should act with respect to others, he replied, "He should cultivate himself so as to ease the lot of others". When asked if there was any one word that could be adopted as a lifelong rule of conduct, he replied, "Is not Sympathy the word? Do not do to others what you would not like yourself". Confucius also valued sagacity, purity, courage, skill and harmony, which "constitute the character of the perfect man" (Confucius c.500 BC/1995: 14–45, 15–23, 14–13).

MacIntyre's interpretation of these differing accounts is that virtues can only be exercised and cultivated in the context of what he terms a *practice*. He gives a very specific definition of what he means by a practice:

> … any coherent and complex form of socially established cooperative human activity through which goods internal to that form of activity are realised in the course of trying to achieve those standards of excellence which are appropriate to, and partially derivative of, that form of activity, with the result that human powers to achieve excellence, and human conceptions of the ends and goods involved are systematically extended.
> (MacIntyre 1981/1985: 187)

Among his examples of a practice are the game of football, chess, architecture, farming, and the enquiries of physics, chemistry, biology and history. MacIntyre makes a distinction between *internal goods* and *external goods*. *Internal goods* need a practice for specification and can only be fully understood by participation in that practice. In the context of a game of chess these might be particular kinds of analytical skill, strategic imagination and competitive intensity. The achievement of such goods involves standards of excellence and obedience to rules. *External goods* are contingently attached to activities and could be achieved in other ways. Examples include prestige, status and money. According to MacIntyre, external goods are always some individual's property and possession, and are achieved in competition resulting in losers as well as winners. In contrast, the achievement of internal goods, though the result of competition to excel, is of benefit for the whole community engaging in the practice.

This analysis allows MacIntyre to postulate a new definition of virtue:

> A virtue is an acquired human quality the possession and exercise of which tends to enable us to achieve those goods which are internal to practices and the lack of which effectively prevents us from achieving such goods. (MacIntyre 1981/1985: 191)

He regards such virtues as defining the relationships between people who share the purposes and standards of a practice. He further regards truthfulness, justice and courage as being prerequisite virtues for any practice. The inclusion here of courage, with its Homeric overtones of an heroic society, appears at first surprising. However, it has special relevance in the present context following from the explanation, "We hold courage to be a virtue because the care and concern for individuals, communities and causes which is so crucial to so much in practices requires the existence of such a virtue" (MacIntyre 1981/1985: 192).

MacIntyre draws attention to several salient features of his analysis. First, that a practice is more than just a set of technical skills. Second, that no practices can survive for an extended time unless they are sustained by *institutions*, of which he gives clubs, laboratories and universities as examples. He regards institutions as characteristically concerned with external goods. They also give practices an historical dimension. Third, as is clear from the initial definition, practices have an aspirational aspect; their goals are subject to continuous extension. Fourth, that if the pursuit of external goals were to become dominant, the concept of the virtues would suffer attrition and even total effacement. In MacIntyre's view, the nature of the world is such that unbridled pursuit of wealth, fame or power leads to loss of truthfulness, justice, courage and other desirable qualities.

Although engineering is not mentioned as an example of a practice in *After Virtue,* it seems clear that it fulfils the requirements of MacIntyre's definition. There are specific *internal goods* in the profession of engineering, in particular those associated with the accurate and rigorous application of scientific knowledge combined with imagination, reason, judgement and experience. These give personal satisfaction as well as providing a framework of excellence on which further activities can be based. The *external goods* of engineering particularly include technological artefacts. Further, considerable economic benefits are generated for the broader societies in which engineering activities take place. The various forms of *institution* that provide continuity and coherence to the practice of engineering include professional engineering associations, universities and commercial enterprises. These institutions can play a key role in promoting the aspirations of individual engineers, so as to *systematically extend* the practice of engineering.

Though *ends* are mentioned in the definition of a practice, it is curious that MacIntyre subsequently writes very little about them. This is maybe because the examples that he develops, such as chess and portrait painting, do not have ends providing substantial physical benefits to the broader community. *The end of engineering may be described as the promotion of human flourishing through contribution to material wellbeing.* Here a cautionary note should also be sounded. MacIntyre noted the dangers of too great a focus on external goods such as wealth, fame or power. In the case of engineering there is also a danger of focussing too greatly on the external goods, technological artefacts, of becoming lost in the labyrinth of technology. In MacIntyre's terminology, technological artefacts should be considered as contingent products, external goods, in the pursuit of the end of human flourishing. *Hence, the present imbalanced prioritisation in engineering of technical ingenuity over helping people may be considered as arising from mistaking the external goods of the practice for the real end of the practice.*

It is also pertinent to inquire about which virtues are appropriate for the practice of engineering. These would be expected to differ from those best suited for an Homeric hero, an Athenian gentleman or a Philadelphian inventor and entrepreneur. MacIntyre defined virtues in terms of the enabling of achievement of goods internal to practices. In the case of engineering this definition would benefit by being broadened to include the enabling of the achievement of ends, especially the end of the promotion of human flourishing through contribution to material wellbeing. MacIntyre has suggested that truthfulness, justice and courage are prerequisite virtues for any practice. An authoritative statement of the further virtues is given in the UK Royal Academy of Engineering's *Statement of Ethical Principles.*

Though the Academy refers to them as principles, they clearly merit consideration as virtues on the understanding presented here (RAE 2007):

> **Accuracy and Rigour** Professional engineers have a duty to ensure that they acquire and use wisely and faithfully the knowledge that is relevant to the engineering skills needed in their work in the service of others.
> **Honesty and Integrity** Professional engineers should adopt the highest standards of professional conduct, openness, fairness and honesty.
> **Respect for Life, Law and the Public Good** Professional Engineers should give due weight to all relevant law, facts and published guidance, and the wider public interest.
> **Responsible Leadership: Listening and Informing** Professional Engineers should aspire to high standards of leadership in the exploitation and the management of technology. They hold a privileged and trusted position in society, and are expected to demonstrate that they are seeking to serve wider society and be sensitive to public concerns.

As part of the introduction of these principles or virtues, the *Statement* contains a cogent and challenging description of what in MacIntyre's terminology might be considered *the practice of engineering*:

> Professional engineers work to enhance the welfare, health and safety of all whilst paying due regard to the environment and the sustainability of resources. They have made personal and professional commitments to enhance the wellbeing of society through the exploitation of knowledge and the management of creative teams. (RAE 2007)

These principles or virtues and the description of the practice of engineering provide a key component in the development of an aspirational engineering ethical ethos. What is most needed in addition is a more positive commitment to prioritising people, to work for the wellbeing of all *individuals* affected by engineering activities. There is a danger that consideration of the "welfare, health and safety of all" or the "wellbeing of society" can be interpreted in terms of a calculus of consequences or even a too narrowly defined contractualism. The terms "all" and "society" need to be interpreted both more specifically, in terms of persons, and also more generally, beyond the boundaries of nation states at which societies tend to draw their limits. *Such terms need to be understood as referring to both human collectivity and the human quality in each individual.* Here concepts such as those of Buber's *I-Thou* relationships or Levinas' *face* can provide vocabularies for further consideration. In expanding our viewpoint we can also learn from religious virtues such as love, benevolence or sympathy, for these have transcended the boundaries of cultures or nation states for millennia.

6.4 Continuity and Coherence: the Life of the Individual Engineer

MacIntyre's virtue based description of practices and institutions provides the basis for a coherent ethical description of engineering as a profession. A further issue that needs to be addressed is how coherence can be achieved for an individual engineer. This is a topic needing specific attention due to the tendency noted in Chap. 4 for employees to bracket their personal ethical values while at work, a susceptibility that seems to be a feature of goal- and end-oriented work environments. The adoption of a genuinely aspirational engineering ethical ethos would be greatly facilitated if this susceptibility could be minimised. Here a simple observation is very pertinent – that a person who genuinely possesses a virtue would be expected to manifest it throughout the range of his or her activities, both professional and personal. This is a very important benefit of virtue-based descriptions of ethics. In contrast, rules and duties tend to apply to rather specific sets of circumstances.

The suggestion is that we need to view our ethical lives as a unity. Indeed, each individual needs to view the whole of his or her life as a unity. Such coherence of professional and personal affairs would give to each of us a greater consistency in ethical choices. In developing such a viewpoint we can refer to the way that MacIntyre and others have drawn attention to the narrative nature of human lives. In trying to understand ourselves we imagine ourselves within a story. We do not view our lives as a series of unconnected episodes, as simply a chronology. Instead we consider the intentions, situations and history of our activities. We have a degree of authorship of our lives. We can set aims and choose directions, though we are of course subject to many unpredictable occurrences including the actions of others. The narrative in which we place ourselves gives an intelligibility to our lives. Our degree of authorship also makes us accountable not just for isolated events but for our overall direction and the cumulative effect of all activities in which we take part.

The consideration of one's life as a narrative unity also provides a valuable means of ensuring that engineers keep the needs of other individuals at the forefront of their awareness and hence avoid becoming lost in the labyrinth of technology. Very few of us would aim to be callous in our personal lives. Indeed, many would be unhappy at the thought of being perceived as being callous even in some small personal action. The idea of a narrative unity can encourage us to maintain consistent aims and actions in both our personal and our professional lives, to live our life in harmony. The concept of self-authorship also enables us to envisage long-term ethical aspirations, or ethical projects, to use a more engineering term. This approach is closely related to the ancient view of ethics as having as an aim the leading of an accomplished *life*.

A narrative view of life is recognised in everyday language. The description of someone having "lost the plot" is used to indicate loss of judgement in a given situation, usually a situation of significant importance in an individual's life. We can recognise this in our own lives and in the lives of others. Indeed, recognition of the need for and the importance of a cumulative narrative in the lives of others can provide a powerful underlying support for providing care. We can become aware of others' stories. As Ricoeur has written, "We tell stories because in the last analysis human lives need and merit being narrated. This remark takes on its full force when we refer to the necessity to save the history of the defeated and the lost" (Ricoeur 1983/1984: 75).

Engineering ethics has been most frequently seen as seeking an answer to questions of the type, "What should I do?", or more specifically, "Which option should I choose?". A virtue approach to engineering ethics, with the virtues defined within the context of the professional practice of engineering, indicates a further important type of question, "What underlying professional characteristics should I cultivate as an engineer?". The challenge of considering one's life as a narrative unity suggests further questions, "Of what story am I a part and how do I wish to develop this story?". However, there is another crucial way in which the approaches differ. The question "What should I do?" can lead to a further series of investigations and questions of the type, "What are the reasons for doing …?". Due to the complexity of engineering, indeed the complexity of life, there are a myriad of such investigations and questions. In the case of consequentialism, contractualism and even, to a large extent, duty based approaches to ethics such questions will in theory have an impersonal answer. However, in practice, at some point the engineer has to make a decision, to make one of the available options his or her own. Virtue ethics in a narrative context to the greatest extent makes clear that the engineer then has to be prepared to make the statement, "*I* am responsible for …".

This chapter has so far made the case for the priority of people, for considering engineering as a practice, for considering an individual's life as a narrative unity, and for the continuity and coherence provided by such considerations in the development of engineering ethics. Ricoeur has provided a challenging statement of the goal of such ethical priorities:

> … to live well with and for others in just institutions and to esteem oneself as the bearer of this wish … (Ricoeur 1990/1992: 352)

This succinct formulation returns attention to the need to consider the nature of and relationship between three types of care: *I* (self care), *I-Thou* (proximate care) and *I-You* (care beyond proximity). Achieving a balance of such commitments is a challenge for all in the modern world, and especially so for professionals such as engineers due to the great reach of engineering activities in both place and time.

6.5 *I*, *I-Thou* and *I-You*: Self Care, Proximate Care, and Care Beyond Proximity

Until modern times, it was only a few very exceptional individuals who could directly influence the lives of those outside their family and local community. In sharp contrast, it is possible today for most people in the more affluent parts of the world to influence the lives of others both close by and far away. Among the ways in which this can be achieved are through international travel, through electronic communications and through international trade. Where we conduct our business and take our holidays, how we interact by email and place information on websites, and our choices of food and other material goods can significantly influence the possibilities for others to flourish. Donations to international relief charities can literally make the difference between life and death for others in distant, impoverished parts of the world. However, we still, as in former times, have responsibilities to care for ourselves and for those closest to us. Thus, one of the most difficult ethical challenges we all face is how to prioritise these different demands. In the extended terminology that we can derive from Buber, how do we prioritise *I*, *I-Thou*, and *I-You* relationships? Alternatively expressed, how do we prioritise self care, proximate care and care beyond proximity?

Achieving a balance of such commitments is especially challenging for engineers as their professional activities may have substantial influence far beyond proximity in both place and time. Further, the special capabilities for promoting well-being which engineers possess bring special responsibilities. Buber noted the especially demanding and enhanced responsibility inherent in the asymmetric relationships which teachers and doctors have with their pupils and patients (Buber 1923/2004: 98–99 (Postscript)). Such an enhanced responsibility to provide care applies likewise to engineers. They have a responsibility *for* those affected by their activities.

A concern for *I*, self care, is a necessary aspect of any ethical viewpoint. A few philosophers have propounded the limiting case of egoism, a concern only with one's self-interest. It still seems to have some lingering influence through occasional, somewhat incoherent, extension of the curiously anthropomorphic concept of the "selfish gene" to social behaviour. However, egoism is not only based on a very inadequate description of human life, but can indeed be regarded as incomplete even in a logical sense (Graham 2004: 17). Nevertheless, as a minimum, if we do not put some effort into taking care of ourselves we will be unlikely to be in a fit state to contribute to taking care of others. Indeed, we rightly regard the ability to progressively provide for oneself, within the framework of society, as a key feature of emergence from childhood and adolescence into adulthood. The idea of considering one's life as a narrative unity also implies a significant degree of self care, not just with single actions but with the entire structure of one's existence. A useful way of expressing such a concern for self in ethics has been succinctly suggested by Buber:

> … everyone should search his own heart, choose his particular way, bring about the unity of his being, begin with himself … What am I to choose my particular way for? The reply is: Not for my own sake. That is why the previous injunction was: to *begin* with oneself. To begin with oneself, not to end with oneself; to start from oneself, but not to aim at oneself; to comprehend oneself, but not to be preoccupied with oneself.
> (Buber 1948/1965: 24–25)

Relationships of the type *I-Thou*, or proximate care, are vital parts of a fulfilled human existence. Thus, *I-Thou* relationships, which characteristically prioritise family, friends and one's local society, should form a very important part of any individual's life narrative. It should be noted that ethical viewpoints based on consequentialism and utilitarianism, as they are fundamentally impersonal in their means of analysis, cannot account for this. Likewise, views of duty of the Kantian type, though they emphasise respect for others, prioritise principles rather than people. Thus, an approach to ethics based on the priority of people, virtue and narrative provides an account of vital aspects of the role of ethics in life that the dominant ethical viewpoints can only inadequately describe. A further important feature of an *I-Thou* relationship is that it is not formed in the expectation of a reciprocal response, it is *beyond reciprocity*, though reciprocity may arise as the relationship develops. This provides another contrast with consequentialism, contractualism and duty, which are based on expected universal patterns of behaviour. This non-reciprocal aspect of priority given to persons is expressed even more strongly by Levinas' strikingly visual notion of *the face* that demands a personal response from us.

The proposal of an *I-You* type of relationship by Cox (1965: 61) was originally intended to describe the appropriate attitude to the many brief encounters that are a feature of modern urban life. Such an *I-You* relationship recognises the uniqueness and integrity of the person encountered but acknowledges that the interaction may be brief and lacking intimacy. There is a great variety of such relationships, but in all there is an aspiration to ensure that they are personal, that the person is not used in a way that leads to degeneration to an *I-It* interaction. Simply put, other people have names. In some cases reciprocity may be maintained, for example in simple commercial exchanges. In others one of the parties may make a response *beyond reciprocity*, for example in helping another in difficulty. In both cases, spatial proximity is experienced, but only for a short time.

It is now proposed that the expression of an aspirational engineering ethical ethos may be aided by extending the *I-You* vocabulary *beyond proximity*, to include a relationship with people who may be distant in place and/or distant in time. Thus, the task of the engineer is to develop technical knowledge and technical activities, the world of *I-It* with its goods of appropriate technical advance, in response to an *I-You* concern for those affected by the technical advance, with the end of the promotion of human flourishing through contribution to material well-being. The people affected by the activities may be located far from the place where the engineering work is conceived and planned. In some cases, they may be far from the place where the engineering artefacts are constructed or even far from the place where the completed, engineered artefacts are located. Again, people may be affected at some future time, perhaps at some distant time. To extend Levinas' notion, the engineer must visualise *the face* of those affected by his or her activities even if they are not physically present, even if they have not been born. This visualisation of *the face* should be accompanied by the internal hearing of the voice who says, "It's me here, please help me!".

A distinction similar to that between proximate care and care beyond proximity has been made by Slote in his development of a "warm agent-based virtue ethics" (Slote 2001). Such an approach to ethics is inspired by the "moral sentimentalist" tradition of Hutcheson, Hume and Martineau. Hume had stressed the importance of virtues, especially benevolence, "it seems undeniable, that nothing can bestow more merit on any human creature than the sentiment of benevolence to a high degree" (Hume 1751/1983: 20). Slote's approach to ethics is based on an attitude of universal benevolence, which he regards as a motive seeking certain ends. This he contrasts with consequentialism that refers to the occurrence of those ends. However, the motive of universal benevolence still has an empirical aspect, for the agent gathers information about the effects of actions. It is an ethic of consequence-sensitivity rather than a calculus of consequences. Another important difference with consequentialism is that "benevolence involves not only the desire to do what is good or best overall for the people one is concerned about, *but also the desire that no one of those people should be hurt or suffer*" (Slote 2001: 28, original italics).

This last important distinction also applies to the approach developed in the present book. However, the present approach also has significant differences in comparison with that of Slote. For example, if it had to be characterised in a similar format, it would be better described as *Thou-, You-* or *face-* focussed, rather than agent-based. The role of virtue is of cultivating the response of care, of *I-Thou, I-You,* of seeing *the face.* Also, distinguishing between *I-Thou/I-You* and *I-It* interactions is crucial in the development of an ethical ethos for a technical profession such as engineering. Again, the *practice* of engineering provides an essential framework for the activities of the individual engineer. Further, the idea of *narrative unity* in ethical life provides an essential cohesion between professional and personal activities.

The need for compassion and generosity in life has been emphasized by Ricoeur, who has used the term *benevolent spontaneity* to indicate a fundamental priority over obedience to duty (1990/1992: 190). Following a discussion of Levinas' work, he describes the full range of motivations to action:

> ... to take an overview of the entire range of attitudes deployed between the two extremes of the summons to responsibility, where the initiative comes from the other, and of sympathy for the suffering of the other, where the initiative comes from the loving self, friendship appearing as a midpoint where the self and the other share equally the same wish to live together. (Ricoeur 1990/1992: 192)

He later writes very favourably of Marcel's term *disponibilité*, availability (Ricoeur 1990/1992: 268)[2]. Ricoeur writes elsewhere of the apparent dichotomy between love (which here might be considered a term appropriate to describe *I-Thou,*

---

[2] Referring to Marcel (1949/1976).

proximate care) and justice (which here might be considered a term appropriate to describe *I-You*, care beyond proximity):

> Rather than confusing them or setting up a pure and simple dichotomy between love and justice, I think a third, difficult way has to be explored, one in which the tension between the two distinct and sometimes opposed claims may be maintained and may even be the occasion for the invention of responsible forms of behaviour. (Ricoeur 1991/1995: 324)

A key term for Ricoeur in the promotion of such new forms of behaviour is *generosity*.

It should be noted that the distinction between *I-Thou* and *I-You* relationships is not necessarily sharp. Due to the complexity of human interactions it is best to allow the possibility of a graduation between these two cases. This can be a useful feature in creating a narrative unity within an individual's personal and professional life. Indeed, a relationship between persons or groups of people may move from *I-You* to *I-Thou*, or the reverse, as time and place change. Further, just as the limitations on a human life limit the number of *I-Thou* relationships it is possible to maintain, so ability to engage in *I-You* relationships may in practice be limited. However, it is always possible to promote this way of thinking and feeling, so as to enhance the extent of others' prioritisation of people – in engineering language, to act as a catalyst.

In seeking to balance our commitments it is vital to acknowledge that there is a limit to the extent that any individual can prioritise ethical behaviour. In an essay on Gandhi, Orwell used the provocative opening words, "Saints should always be judged guilty until they are proved innocent …" (Orwell 1949/2000: 459). There can be apparent in "sainthood" a certain lack of a full range of human values. Gandhi, for example, seems to have avoided close friendships. In the same essay, Orwell wrote, "Many people genuinely do not wish to be saints, and it is probable that some who achieve or aspire to sainthood have never felt much temptation to be human beings". We each have to live within our own capabilities and potential. Buber has written very succinctly on this issue in his report of the words of Rabbi Zusya shortly before his death, "In the world to come I shall not be asked: 'Why were you not Moses?' I shall be asked: 'Why were you not Zusya?'" (Buber 1948/1965: 10).

For professionals such as engineers the need to balance commitments occurs in (at least) two ways. There is firstly the need to balance *I*, *I-Thou* and *I-You* in an overall coherent manner in personal and professional life. There is secondly a need to balance *I-You* commitments to different *Yous* in professional activities. Such balances require sensitivity, including sensitivity to consequences. For neither way is a *calculation* possible. For each way a *decision* has to be made. Making a decision can be difficult and merits much effort. Such decisions are always potential bearers of guilt, as Derrida has explained:

> Guilt is inherent in responsibility because responsibility is always unequal to itself: one is never responsible enough. One is never responsible enough because one is finite …
> I cannot respond to the call, the request, the obligation or even the love of another without sacrificing the other other, the other others … I am responsible to any one (that is to say to any other) only by failing in my responsibilities to all the others, to the ethical or political generality. (Derrida 1992/1995: 50, 68, 70)

The potential for guilt also makes clear that when we make a decision affecting others we are also making a decision about ourselves.

Most often the focus in such ethical decision making has been on the resolution of specific ethical quandaries. However, an essential antecedent approach is to adopt a positive way of being the very nature of which helps to minimise the occurrence of such difficult situations. The outline of such a way, an aspirational ethical ethos for engineering, has been the theme of the present chapter. As both engineering and ethics are practical, such an engineering ethic should be doubly expected to have practical outcomes. A description of some important outcomes of the aspirational ethos outlined is the subject of the next chapter.

References

Anscombe GEM (1997) Modern moral philosophy. In: Virtue ethics, Crisp R and Slote M (eds), Oxford University Press, Oxford, p. 26–44. Originally published 1958

Buber M (2004) I and Thou, Continuum, London. Originally published as Ich und Du, Berlin: Shocken Verlag, 1923

Buber M (1965) The way of man, Routledge, London. Originally published 1948

Confucius (1995) The analects, Dover Publications, New York. Original attribution c. 500 BC

Cox H (1965) The secular city, Penguin Books, London. Originally published by Macmillan, New York

Derrida J (1995) The gift of death, University of Chicago Press, Chicago. Originally published as Donner la mort in L'éthique du don, Jacques Derrida et la pensée du don, Métailié-Transition, Paris, 1992

Graham G (2004) Eight theories of ethics, Routledge, London

Hume D (1983) An enquiry concerning the principles of morals, Hackett Publishing Company, Indianapolis. First published 1751

Keown D (2005) Buddhist ethics: a very short introduction, Oxford University Press, Oxford

Korsgaard CM (1996) The sources of normativity, Cambridge University Press, Cambridge

Levinas E (1969) Totality and infinity, Dudesque University Press, Pittsburgh. Originally published as Totalité et infini, Martinus Nijhoff, The Hague, 1961

Levinas E (1981) Otherwise than being, Dudesque University Press, Pittsburgh. Originally published as Autrement qu'être, Martinus Nijhoff, The Hague, 1974

Mackie, JM (1977) Ethics: inventing right and wrong, Penguin Books, London

MacIntyre A (1985) After virtue, 2nd edition, Duckworth, London. First published 1981

Marcel G (1976) Being and having: an existentialist diary, Peter Smith, Gloucester. Originally published 1949

Orwell G (2000) Reflections on Gandhi, in "Essays", new edition, Penguin Books, London. Originally published 1949

Ricoeur P (1984) Time and narrative, vol 1, University of Chicago Press, Chicago. Originally published as Temps et récit, Editions du Seuil, Paris, 1983
Ricoeur P (1992) Oneself as another, University of Chicago Press, Chicago. Originally published as Soi-même comme un autre, Editions du Seuil, Paris, 1990
Ricoeur P (1995) Figuring the sacred, Augsburg Fortress, Philadelphia. Chapter originally published 1991
Royal Academy of Engineering (2007) Statement of ethical principles, RAE, London
Slote M (2001) Morals from motives, Oxford University Press, Oxford
Stocker M (1997) The schizophrenia of modern ethical theories. In: Virtue ethics, Crisp R and Slote M (eds), Oxford University Press, Oxford, p. 66–78. Originally published 1976

Chapter 7
Practical Outcomes

The outcomes of both engineering and ethics are practical. Hence, any analysis of engineering ethics should be doubly expected to have practical outcomes. The purpose of the present chapter is to consider some of the outcomes of the analysis developed. These outcomes relate to the promotion of an aspirational ethical ethos for the profession rather than to specific moral quandaries. It may be helpful to first briefly summarise some of the main points of the analysis.

7.1 Overview of Analysis

At its best, engineering changes the world for the benefit of humanity. However, the great technical successes of engineering and the enormous satisfaction that engineers can gain from the purely technical aspects of their work lead to a danger. The danger is that engineers forget that technical ingenuity in itself is not the goal of engineering. They become lost in the labyrinth of technology. In their commitment to the advancement of technology they forget that the goal of engineering is the promotion of human flourishing through contribution to material wellbeing.

The unfortunate tendency to prioritise technical ingenuity over helping people has a pervasive presence in modern engineering. This tendency has two main types of undesirable outcome. Firstly, the development and application of complex but inappropriate technology. A very significant example is military equipment and weapons. Enormous quantities of engineering resources and skills are committed to the design, manufacture and use of such technologies. The use of many of the weapons produced is in contravention of international conventions and treaties. Secondly, technology that could alleviate great human suffering is available but is not being applied as extensively or as rapidly as it could be. A major example is the lack of provision of clean water and effective sanitation for many of the world's population. As a result, 1.1 billion people do not have safe drinking water, 2.4 billion people have no provision for sanitation, and 6,000 people, mostly children under the age of five, die *every day* from water related diseases.

Ethical analysis of engineering has been limited compared to that of other key professions. Where such analysis has been carried out, there has been a tendency for engineers to adopt consequentialism as a default position. The appeal of this position to engineers is related to its apparent simplicity and quasi-empirical basis combined with an inadequate appreciation of its philosophical limitations. Contractualism is also a view that has been widely applied in analysis of engineering ethics. Some engineers would view this as a default justification of their activities: if some aspect of their work is not explicitly legally forbidden then it is regarded as acceptable. The widespread development of weapons that contravene international law shows that even this minimalist view is not always adhered to. More generally, contractualism can be a justification for ethical mediocrity and acceptance of present minimum norms. An emphasis on duty, deriving in philosophical terms from Kant, has also played a significant role in the development of engineering ethics, especially as a background to the formulation of ethical codes. The respect for persons inherent in a duty approach has much to offer engineering ethics. However, the often dense argumentation associated with philosophical discussions of this approach is a barrier to its more widespread adoption by engineers. Virtue based approaches have generally had less influence on engineering ethics than the other traditional approaches despite the great sophistication of Aristotelian ethics and the recent resurgence in philosophical interest in virtue ethics.

Engineering ethics can learn from the more advanced ethical development in other professions, especially medicine and business. Medicine is the profession that has had the longest and most intense interest in ethical concerns. Modern medical ethics differs in two striking ways from the generally adopted approaches to engineering ethics. Firstly, there is a great emphasis on the *individual*, the patient, affected by the doctor's actions. Secondly, the doctor is emphatically *personally* accountable for decisions. As a result, for example, doctors are authoritatively advised that they should not knowingly use their skills and knowledge for weapons development and that their primary duty to avoid harm rises above duties such as those to contribute to national security. A key feature of the difference between the work of doctors and engineers is proximity. Those affected by a doctor's work are usually in his or her immediate proximity whereas those affected by an engineer's work are frequently distant from him or her in both place and time.

Insights from business ethics are also of great relevance to engineering ethics as engineers are frequently employed in commercial enterprises, often large organisations. There is a strong international interest in business ethics at many levels. Thus, the United Nations prioritises the responsibility of business to respect human rights, fair labour practices, environmental concern and prevention of corruption. Some major multinational companies have Codes of Conduct for staff that are both comprehensive and subtle. Philosophical approaches to business ethics are well developed. Two concerns that have become apparent in the study of business ethics need specific attention in the development of an effective approach to engineering ethics. Firstly, a tendency for individuals to bracket their personal ethical values when at work, a suppression of ethical responsibility that seems to be a feature of goal- and end-oriented work environments. Secondly, social science studies of cognitive moral development which suggest that employees often function at a level of minimally acceptable ethical behaviour rather than aspiring to the highest achievable levels.

An aspirational engineering ethical ethos needs to have the capability of promoting the balanced development of the profession. It needs to overcome the limitations of the traditional ethical views and to learn from the more advanced analysis of ethics in other professions. The aspirational ethic outlined takes as a basis a description of an engineer's activities in terms of *I-It, I-Thou* and *I-You* interactions. *I-It* describes the essential technical part of an engineer's work. This results in both *internal goods*, including standards of technical excellence and the satisfaction arising from personal accomplishment, and *external goods*, including engineered artefacts and wealth. The description of engineered artefacts as *goods* allows such physical technological accomplishments to be distinguished from the *end* or *goal* of engineering activity, which is the promotion of human flourishing through contribution to material wellbeing.

I-Thou and *I-You* express the priority that should be given to individual people. An *I-Thou* relationship requires proximity, so that most engineering activities will involve *I-You* interactions, based on care but lacking personal intimacy. *I-Thou* relationships, and even more so *I-You* relationships, require a commitment beyond reciprocity. An *I-You* relationship is additionally beyond proximity. The ethical movement beyond reciprocity may be forcefully expressed by the notion of *the face* that speaks directly, "It's me here, please help me!", and demands a personal response. In the ethical movement beyond proximity the imagination must be used to visualise *the face* of the affected person at a distance in place and time.

The ability to give ethical priority to people is facilitated by cultivation of *virtues*. In general, the appropriate virtues will depend on context. At a professional level, the continuity and coherence of the application of this approach is promoted by consideration of the appropriate virtues in terms of a *practice* supported by *institutions*. It has been suggested that truthfulness, justice and courage are prerequisite virtues for any practice. For the particular case of engineering, the appropriate virtues (originally referred to as principles) have been categorised as: accuracy and rigour; honesty and integrity; respect for life, law and the public good; and responsible leadership, listening and informing.

At a personal level, the continuity and consistency of an ethical approach giving priority to other people may be promoted by an individual considering his or her life as a *narrative unity*. This helps an individual to aspire to ethical behaviour in both private and professional life. In particular, a narrative view allows an individual to reflectively develop a balance between *I*, *I-Thou* and *I-You*, or alternatively expressed, self care, proximate care and care beyond proximity. These different types of caring are mutually supportive. Each prioritises persons in a consequence-sensitive manner. Achieving a balance implies that we each live within our capabilities and potential, also being mindful that there are important human values apart from the purely ethical.

Some outcomes of this approach will now be outlined. The approach suggests changes in the emphasis of engineering education, changes in the role of engineering institutions, changes to industrial work practices and the need to position engineering in the public and intellectual mainstreams. As engineering transcends national boundaries, an example of how engineering may be fully incorporated in international political initiatives will be given. As engineering transcends cultural boundaries, the approach adopted, which has been based on Western conceptual categories, will be compared to that of an Eastern conceptual scheme. Most importantly of all, it is hoped that the approach will stimulate changes in attitude so as to promote the *personal* ethical responsibility of every engineer in his or her professional activities.

7.2 Engineering Education

Engineers enter the profession through university level education. If engineering is to continue contributing to human wellbeing it is essential to recruit able and highly motivated young people with a desire to help others. However, despite there being a strong interest among young people about how they can help others and protect the environment, there is likely to be a shortage of graduate engineers in the future unless recruitment to university courses is increased (RAE 2007). Engineering is not viewed by young people as being a caring profession that can make a real difference in providing the basis for human flourishing and in helping to “save the planet”. The analysis presented here points to three changes in education that would encourage young people to study engineering and subsequently follow careers that prioritised people.

The first change that should take place is in the presentation of the purpose of engineering to potential students. As examples, consider the following introductory statements of purpose. The first is a typical and conventional statement:

> Chemical engineering is the design, development and management of a wide and varied spectrum of industrial processes. Chemical engineers call on their knowledge of mathematics and science, particularly chemistry, to solve technical problems safely, economically and in a sustainable manner.

This statement emphasises the core technical content of one of the branches of engineering. Technical content is, of course, an essential part of engineering. However, such an introductory statement is rather unlikely to attract the attention and stimulate the imagination of an eighteen-year-old. Fortunately, the presentation of engineering to young people is beginning to be carried out in a more enlightened manner as illustrated by the following example:

> Biochemical engineering is concerned with linking discoveries in the life sciences to real outcomes in terms of quality of life. It is a subject of high professional standing and its contribution to providing, for example, the new generation of medicines – recombinant vaccines for cancer, cultured human tissue for repair, genes for therapy of children with cystic fibrosis, new antibiotics to do battle with drug resistant organisms – will be decisive in this century. As a university degree it is exciting and provides a superb foundation upon which to build your future career. (Hoare 2007)

This statement immediately introduces the purpose of the discipline, "real outcomes in terms of the quality of life", with specific and readily understood examples. The statement gives a clear impression that studying this subject could be a stimulating and fulfilling enterprise.

Recruitment to engineering courses should give greater emphasis to engineering as a profession that aims to benefit humanity through the appropriate advancement and application of technology. This would very likely also help change the major gender imbalance that exists in the profession. Professional engineers are overwhelmingly male in many countries – we are in effect only recruiting from about half of the population. This suggestion is supported by evidence from degree programmes in the relatively new discipline of medical engineering, in which about half of students are female.

The second change that should take place in engineering education is in the extent and emphasis given to engineering ethics in the teaching of degree courses. There is already a welcome and increasing awareness that ethics should be a part of university engineering studies and many degree programmes now have an ethics component. This may be described in a way that does not use the term "ethics", such as "professional responsibility". Substantial effort has been invested in developing ways in which engineering ethics teaching can be incorporated in curricula (RAE 2008). There has been debate about the benefits and disadvantages of teaching engineering ethics in different ways, for example getting the balance right between teaching of fundamental ethical principles and case studies.

These very positive developments need to acquire an additional focus on engineering ethics as an *inherent* part of the *overall aims* of professional activities. For example, case studies are a valuable teaching method, but an over-reliance on such an approach can wrongly suggest that ethics is about quandaries, that engineering ethics is something that only merits consideration when things go wrong. It is also important that case studies are formulated to deal genuinely with the effects of engineering activities on *people* – there can be a tendency for examples chosen to be more related to legal than ethical considerations. Teaching of engineering ethics needs especially to emphasise the personal responsibility of each engineer for the effects of his or her activities on others, including where appropriate the imaginative visualisation of consequences for individuals who are remote in place and time.

Once the significance of ethics for the overall aims of engineering is acknowledged, it becomes apparent that one of the most important ethical decisions an engineer makes may well be the choice of first job leading to his or her career path, especially regarding the type of industry or organisation chosen. The provision of more encompassing guidance in this choice is the third change that should take place in engineering education. Engineering programmes should provide students with an understanding of how their chosen discipline can contribute most effectively to the wellbeing of others. Students will be aware from the media of problems such as increasing militarisation, food and water shortages, epidemic disease and climate change. However, they may fail to appreciate how effective use of their skills can make significant contributions to solving these problems and hence alleviate great human suffering. They may also fail to appreciate that their skills are highly transferable and hence applicable to many pressing problems outside of the traditional employment paths for their discipline.

Students need to be encouraged to understand the personal responsibility that they have for the outcomes of their professional activities and how the imaginative use of their skills can bring themselves and others many benefits. Professional engineers are fortunate in being remunerated at levels that allow them to live comfortably virtually irrespective of the particular application to which they put their skills. They are therefore relatively free to consider ethical implications of their chosen career path. They need to be encouraged to chose creatively. As an illustration it is worth considering again the issues identified in Chap. 2. Graduates from many of the sub-divisions in engineering have the skills necessary to help provide water and sanitation in underprivileged parts of the world. If they are made aware of such problems as part of their studies, and of their ability to contribute to solutions, they may well choose to consider such work as part of their career. Again, some of the sub-divisions of engineering are particularly associated with military work. Indeed, a student may well have chosen to study such a sub-division because of a desire to contribute to international security. If such students are to make informed career choices it is important that as part of their studies they are made aware of the scope of international conventions and treaties on weapons development and use, and also that security should be considered in a much wider context than the purely military. They may then consider it more strategic to use their engineering skills in developing a career that helps prevent conflict, such as through provision of energy, water or other relief of poverty.

7.3 Engineering Institutions

In MacIntyre's analysis, institutions play an important role in sustaining the continuity and coherence of practices such as engineering. Institutions exist at many levels. The institutions that at present have the greatest influence on the working lives of engineers are the national professional institutions. In the UK these are the thirty-five engineering institutions licensed by the Engineering Council, such as the Institution of Civil Engineers and the Institution of Chemical Engineers. Additionally and independently, the Royal Academy of Engineering, a national academy, brings together the country's most eminent engineers from all sub-disciplines.

The first way in which these institutions can promote high aspirations in their members is in their overall projection of their sub-disciplines. A good example is the following description of civil engineering:

> … a great art, on which the wealth and well being of the whole of society depends. Its essential feature, as distinct from science and the arts, is the exercise of imagination to fashion the products, processes and people needed to create a sustainable physical and natural built environment. It requires a broad understanding of scientific principles, knowledge of materials and the art of analysis and synthesis. It also requires research, team working, leadership and business skills. (ICE 2008)

More specifically, these organisations have promoted engineering ethics through the development of professional codes. There has been a recent strong resurgence of interest in the formulation of such codes and it is notable that they are becoming more aspirational in content. For example, the Engineering Council's guidelines for the codes of institutions opened with the statements, "Prevent avoidable danger to health or safety. Prevent avoidable adverse impact on the environment" (see Chap. 3), whereas the more recent Royal Academy of Engineering's *Statement of Ethical Principles* (developed in collaboration with the Engineering Council) states more positively, "Professional Engineers work to enhance the welfare, health and safety of all whilst paying due regard to the environment and the sustainability of resources" (see Chap. 6).

This movement toward aspirational formulations by institutions is very welcome. It is also worth considering how it can be further encouraged. One way may be through a re-evaluation of our concepts of justice. Most contemporary societies have fortunately moved beyond a concept of justice as vengeance, "an eye for an eye and a tooth for a tooth". The concept of justice which is most commonly referred to generally in society appears to be closely related to the "golden rule", "Do to others as you would have them do to you", a concept of equivalence which has historical formulations in many cultures. However, an aspirational ethic needs to move beyond reciprocity. If institutions have the role of sustaining the systematic extension of ends and goods, as MacIntyre has suggested, then this movement beyond reciprocity needs to be increasingly incorporated in their ethical formulations. Such a process has been suggested by Ricoeur as the conclusion of a study of the nature of justice:

> I would say that the tenacious incorporation, step by step, of a supplementary degree of compassion and generosity in all our codes – including our penal codes and our codes of social justice – constitutes a perfectly reasonable task, however difficult and interminable it may be. (Ricoeur 1995: 329)

This is certainly a reasonable task and a worthy aspiration for engineering codes of ethics.

More recently, several smaller organisations have been formed which specifically aim to take people-centred approaches to engineering that reach beyond technical issues. For example, Engineers Against Poverty is a specialist organisation working in the field of international development (EAP 2008). It aims to unlock the skills and resources of the private sector to secure social improvements through mechanisms that also deliver commercial advantages to the companies involved. Its work is based on the development of multi-sector partnerships between engineering services contractors, government and civil society organisations. A key feature of this approach is that the under-privileged people should be active participants in the activities rather than just the recipients of benefits. In this way the sustainability of improvements is facilitated. This and similar organisations provide very effective ways of increasing the application of appropriate engineering technology and they merit wide support from professional engineers. Support at present seems to come mainly from civil engineers, though most of the other subdisciplines in engineering could make significant contributions.

There are also several organisations whose aims include the ethical development of engineering even though this may not be explicitly expressed as such. One of the most significant in the UK is Scientists for Social Responsibility, an individual membership organisation that aims to promote ethical science, design and technology (SSP 2008). It carries out research, education and lobbying centred around the military, environmental and political aspects of science, design and technology, also providing a support network for ethically-concerned professionals in these fields. Its publications provide advice on ethical aspects of careers in science and technology. Such organisations also merit wide support from professional engineers. Scientists for Social Responsibility is affiliated to the International Network of Engineers and Scientists for Global Responsibility, a network of about 100 similar organisations in about 40 countries.

A significant development in the US is the emergence of the concept of *peace engineering,* an approach especially associated with P Aarne Vesilind. Peace engineering is envisaged as "the proactive use of engineering skills to promote a peaceful and just existence for all people" (Vesilind 2005: 9). It is seen as a new kind of engineering rooted in the greater ideals and aspirations of engineering as a service to all of humanity. This use of engineering skills to promote peace is considered to be a third kind of engineering, additional to the dichotomy of military engineering and civil engineering (broadly understood). Such an approach questions the morality of the involvement of engineers in weapons research. It places great emphasis on the importance of international and intercultural understanding, including the use of engineering to promote justice and sustainability. Thus, its activities include responding to international emergencies, working in the Peace Corps, working in international development and working to change the nature of engineering education.

Such peace engineering is not glamorous and does not use the latest technology, but it is a highly effective means of preventing conflict through the priority it gives to helping people in extremely vulnerable situations. It recognises that the sources of conflicts are many, that they are frequently related to poverty, and seeks to address them at the local level. Such work requires intense personal commitment from the individual engineer working in the developing country.

The concept of peace engineering is consonant with the development of the aspirational engineering ethos proposed in the present book. However, the present book also emphasises the importance of an aspirational ethos for all engineering activities, without distinguishing different kinds. It may be best to consider the approaches as complementary. Which is the most fruitfully applied may depend on specific cultural circumstances.

In the case of medicine and business, there are influential international institutions and initiatives promoted by the United Nations, respectively the World Health Organisation and the Global Compact. The World Medical Association also plays an important role in promoting the highest possible standards of ethical behaviour and care by doctors. Some aspects of the work of these institutions were described in Chap. 4. In engineering, the World Federation of Engineering Organisations seeks to provide international leadership for the profession. However, its influence on international initiatives appears at present to be modest. It does not have a high public profile, indeed, it is not well-known even among professional engineers. Nevertheless, it shows a strong commitment to ethical engineering (WFEO 2001) and a strengthening of its professional role and public profile would provide major benefits.

International interaction in engineering is more typically promoted through the many international academic and professional societies that aim to develop specialised technical areas. These seek to promote excellence, mostly through the organisation of international technical conferences. However, in view of the enormous influence which engineering has on human wellbeing, the potential that it has for causing great human suffering if misused, and the need to promote its aspirational development, there is a strong case for international institutions such as the United Nations taking a more direct involvement in the discipline. In the absence of such involvement, it is not clear that the essential role of engineering in the accomplishment of such important targets as the Millennium Development Goals will be fully exploited. A specific response to such an omission will be suggested in a later section of the present chapter.

7.4 Industry and Work Practices

Some large engineering companies already have an approach to professional ethics that is sophisticated and subtle. Such approaches deserve praise and promotion as exemplars. In other cases a company-promoted approach to engineering ethics based on a minimally acceptable level of compliance seems to be adopted: involvement with projects which are recognised as being ethically dubious may be rationalised on grounds such as that another company would in any case carry out the work or that the activity is not specifically illegal. The latter approach is reprehensible and its continued existence shows the importance of the need for the promotion of more enlightened views of engineering ethics at the level of whole organisations as well as at the level of individuals.

Two concerns that require special attention are the suppression of personal ethical responsibility in goal-driven work environments and the tendency of employees to function at a level of ethical responsibility which meets only the minimum requirements of their colleagues' and employers' expectations, which may not be high. Encouragement of the concept of a narrative unity in an individual's life can be part of a response to both of these concerns. This is already apparent in the BP Code of Conduct considered in Chap. 4 with its emphasis on an employee's personal responsibility, its commitment to support of an employee's personal priorities and its suggested clarifying question, "What would others think of your action – your manager, colleagues or family?". The reference to family in this question provides a pertinent caution against bracketing personal ethical values at work and a reminder to maintain a whole-life attitude to conduct. Indeed, the promotion of coherence across personal and professional ethical viewpoints and behaviour may be the single most effective means of ensuring the adoption of an aspirational ethical ethos in engineering.

The analysis that has been presented also offers an accessible approach to encouraging engineers to aspire to the highest achievable ethical levels. The highest level proposed by Kohlberg was that of conscience or principle orientation. He identified conscience as engendering mutual trust and respect, with principle orientation involving appeal to logical universality and consistency (see Chap. 4). In general, engineers have a relatively inadequate understanding of the traditional principle based approaches to philosophical ethics. This is to a significant extent due to the rather complex formulation of these approaches. They demand considerable intellectual commitment and time for understanding. However, the basis of the present analysis in the priority of people as identified by Buber and Levinas provides a much more direct approach. In this context, it is worth noting again that Korsgaard reached a similar conclusion on the importance of the priority of others on the basis of a sophisticated and elaborate Kantian-duty based argumentation of a type that only the most ethically committed of engineers are likely to follow. Indeed, a response to her argumentation has suggested that she consider Levinas' work (Williams 1996: 216). The application of MacIntyre's ideas of practices, virtues, institutions, goods and ends is also of a type which can be readily appreciated by engineers as it corresponds to key features of professional life with which they are familiar. There is, therefore, a good prospect that the overall analysis can promote a move to a higher level in the ethical thinking of engineers.

In the comparison of engineering with medicine it was noted that one of the main differences was the greater degree of proximity in both place and time of a doctor and patient than of an engineer and those affected by his or her activities. This is undoubtedly a very important factor in the emphasis on the priority of the affected individual and of personal accountability for professional decisions in medical ethics. Greater emphases on the priority of people and on personal responsibility are key features of the aspirational engineering ethical ethos that has been proposed. The comparison with medicine suggests that the realisation of this ethos may be facilitated if engineering activities could be restructured to increase real or perceived proximity between engineers and those affected by their work. The possibilities for changes in work practices depend on the size, nature and business area of commercial enterprises. The range of possibilities is great and the comments here will be restricted to some observations on relevant differences between large and small enterprises.

A characteristic feature of large enterprises is that each employee's role is for the most part specialised and only one component of a much larger team effort that eventually leads to a product or process. Training and good communications are therefore very important if each individual employee is to appreciate fully the overall effect of the team effort on others. Another important means of increasing real and perceived proximity is to ensure that an individual's career progression involves some participation in those parts of the company's work which are closest to the customers. For companies operating internationally it is also important to educate employees about the different cultures that they are likely to encounter so that they may respond sensitively whilst working in differing societies.

In the UK and the rest of Europe there has been a decline in large heavy industries and a growth in more specialised companies, small and medium enterprises (SMEs). This is leading to changing employment opportunities for engineers. For example, chemical engineers are now less likely to find employment in petrochemical industries and are increasingly likely to work in businesses operating on a more modest scale in food processing or environmental protection. This change of scale in itself leads to a closer proximity, in both place and time, to customers. A smaller scale also often naturally involves an individual engineer being engaged in many roles either simultaneously or over short periods of time, including those that are closest to customers. These features are all likely to help minimisation of ethical bracketing and to increase awareness of personal responsibility.

It is worth noting that companies producing military equipment and weapons use procedures opposite to those suggested here. This begins with the manipulation of language even in descriptions of core business, so a common designation of military equipment and weapons enterprises is "defence and aerospace", where even the military connotations of defence are diluted by the addition of aerospace. The use of the products of these companies is always described euphemistically. For example, a bombing raid is described as a "surgical strike", in an attempt to allude to a beneficial medical procedure, or dead civilians are described as "collateral damage", ignoring their humanity. The latter is an example of an attempt to limit thought by the use of a terminology that it is impossible to visualise (Orwell 1946/2000: 348). The work of individual engineers in such companies will be described in similarly euphemistic terms, so that its real purpose is not clearly apparent. Further, there will be a strategy of minimising the individual's appreciation of the overall purpose of their work so as to curtail even the imagination of proximity to those affected by the final products. Borges has perceptively described a similar phenomenon in his short story about a woman, Emma Zunz, on her way to commit a murder:

> Paradoxically, her weariness turned to strength, for it forced her to concentrate on the details of her mission and masked from her its true nature and final purpose. (Borges 1949/1998: 218)

Emma Zunz believed that the murder would restore justice.

7.5 Positioning Engineering in the Public and Intellectual Mainstreams

Discussions of the practice of medicine and business form part of the mainstreams of public and intellectual debate in Western countries. For example, there is a vigorous and public debate about medical ethics and the prioritisation of medical resources. There is a correspondingly energetic debate about business. The public interest in medicine is also shown by the popularity of medical dramas in television and film. The public and media interest in engineering appears to be more muted.

There has not always been such indifference. At the beginning of the modern mass media era, the work of one of the greatest figures in the development of silent films, Buster Keaton, showed his fascination with the engineering of his time. Trains feature in many of his films and are central to his masterpiece *The General* (1927), that title being the name of a locomotive. Many of his films feature boats, from family vessels to ocean liners: (*The Boat* (1921), *The Navigator* (1924) and *Steamboat*

Bill Jr (1928)). *One Week* (1920) is based around an innovative method of house construction and the plot of *The Electric House* (1922) arises from the work he carries out after mistakenly graduating as an electrical engineer. Film was, of course, an engineering success in itself and the development of technology is essential to the plot of *The Cameraman* (1928). However, Keaton is unique in basing so much of the core of his work around engineering. Other films, outside of science fiction, have only occasionally featured engineering centrally: recent examples include *Apollo 13* (1995), *Titanic* (1997) and *The Aviator* (2004). Films both form and reflect public values and perceptions. However, regardless of whether formation or reflection is dominant, it might be concluded that this general sparsity shows that engineering does not feature highly in the public's interest and imagination.

An exception to this general observation, as the films *Apollo 13* and *Titanic* indicate, is the public interest in and memory of accidents and disasters. It is certainly important to be concerned about events such as the Windscale fire, the Chernobyl explosion, the explosion in the chemical plant at Flixborough, or the Piper Alpha explosion, the world's worst offshore oil disaster. The loss of the Titanic is just one of many ship losses, shipping vessels over the whole range of sizes being the engineered products which are most likely to be unable to cope with the conditions which they experience. Accidents and disasters also feature strongly in the memory and imagination of engineers. Analysis of such events forms the core content of some books about engineering ethics. However, too great an emphasis on such failure can give an unbalanced view of engineering.

Another feature of the public perception of engineering is that it often focuses on protests against engineering developments. Newspaper headlines such as, "Anger at Heathrow expansion plans", relating to concerns about noise, pollution and damage to homes, "Trident protests", concerning weapons development, and "Arrests at gas pipeline protest", regarding the construction of a natural gas pipeline in a National Park, are the forms in which engineering normally reaches the front pages of newspapers. Reports of major engineering achievements also appear, but they do not seem to have the same impact or longevity as the negative accounts.

Engineering and engineers do not feature strongly in the world's great literature. Indeed, the Tay Bridge Disaster of 1879 even famously formed the subject matter of a poem by William McGonogall, widely regarded as one of the English language's worse poets, which concludes with the lines (McGonogall 1890/2004):

> I must now conclude my lay
> By telling the world fearlessly without the least dismay,
> That your central girders would not have given way,
> At least many sensible men do say,
> Had they been supported on each side with buttresses,
> At least many sensible men confesses,
> For the stronger we our houses do build,
> The less chance we have of being killed.

This is certainly better engineering than it is poetry. More recently, several of Tom Lehrer's memorable songs have concerned the amoral or immoral development and application of engineering. For example (Lehrer 1965):

> When the rockets go up who cares where they come down?
> "That's not my department", says Wernher von Braun.

Fortunately for engineering, von Braun is almost always described as a rocket *scientist.*

A noteworthy exception to this lacuna in serious literature is the novel *The Spire* by William Golding (1964), a powerful evocation of struggle in its many forms set in the context of a great construction project. The Dean of a medieval cathedral decides to add a four hundred foot spire to the building. He pushes forward with the project despite huge practical obstacles and the advice of those around him. Golding has described the book as being about "the human cost of building the spire" (Golding 1988: 166). The Dean's observations on those carrying out the work are perceptive:

> He saw the busy shapes of men who did as they were told but did not know what they were doing. (Golding 1964: 69)

This comment is still pertinent to some aspects of modern engineering. The project is brought to completion, but at a huge personal cost to the central characters in the story. The Dean eventually realises the folly of his ambition:

> Imagine it. I thought I was doing a great work; and all I was doing was bringing ruin and breeding hate. (Golding 1964: 209)

The novel is a finely wrought reminder of the need to place human wellbeing at the forefront of engineering endeavours. The complexity of the intention and interpretation presented also presents a challenge to engineers to engage in similarly honest reflection about both positive and negative aspects of their activities.

The move toward the successful adoption of an aspirational ethical ethos for engineering needs to engage the active participation of the public and intellectual mainstreams. This engagement needs to be carried out at several levels. The first and most important level is the formulation, in the most rigorous manner possible, of an expression of the need for and the benefits of engineering. In particular, the contribution of engineering to the promotion of human flourishing through contribution to material wellbeing needs to be clearly expressed. At present engineering tends too often to be perceived as a somewhat nerdy technical discipline. In this context, engineers could learn lessons from the environmental movement. This has been able to build on the work of philosophers such as Arne Naess, who through works like *Ecology, Community and Lifestyle* (Naess 1989) have been able to provide a strong intellectual evaluation of the need for giving priority to the solving of environmental problems. Such works inspired the *deep ecology movement* that goes beyond environmental science to consider values that can be developed into an environmental ethics. In essence, deep ecology emphasises the intrinsic importance of nature for the humanity of man. It might be suggested that the time is now right for a *deep engineering movement* that could emphasise the intrinsic importance of engineering to the promotion of human flourishing through contribution to material wellbeing. Part of this importance lies in the potential for engineering to solve environmental problems, and indeed, more positively, to care for the environment. It is hoped that the present work can make a contribution to this new understanding.

There could also be a benefit in complementing such intellectual analysis with institutional reform, a suggestion that moves engineering ethics towards an uncertain boundary with what might be termed engineering politics. At a national level, it may again be possible for engineering to learn from practices in the medical profession. For example, appropriate practice in fertility treatment and embryo research, which is a very contentious subject ethically, is independently regulated in the UK by the Human Fertilisation and Embryology Authority. This authority licenses and monitors work such as in-vitro fertilisation and human embryo research. It also provides detailed information on such issues for patients, professionals and government. The formation of an authority to license and monitor major engineering projects could be a helpful means of replacing the current heated protests with more rational discussion. Such a body could have as its task the provision of reasoned evaluation that goes beyond narrow and immediate concerns. It could provide a means of discussing and evaluating the needs and desires of people, engineering enterprises and government in an aspirational manner.

As a concluding comment on the public perception of engineering and of engineering in the intellectual mainstream, it should be noted that although there is a Nobel Prize for literature there is no such prize for engineering. However, engineering could still benefit substantially from the Nobel Foundation if someone could emulate Golding's achievement by writing a prize-wining novel about a more modern aspect of the subject!

7.6 An Aspirational Role for Engineering in International Political Initiatives

Initiative is also possible at an international level. Mention was made in Chap. 2 of the United Nations *Declaration and Programme of Action on a Culture of Peace* (UN 1999). Such a *Culture* is characterised by relationships between individuals, and social groupings of all sizes, based on honesty, fairness, openness and goodwill. It was noted that engineers were not included in the official list of professionals who could contribute to such a *Culture of Peace*, perhaps because of their perceived close alignment to the military, perhaps due to simple error, perhaps for some other reason. Depending on the reason, the omission shows either mistrust of, or ignorance about, the potential contribution of engineering at the highest political level. Whatever the explanation, it is timely and necessary to remedy the omission, for the beneficial contribution of the development of an aspirational engineering ethos is likely to be greatly enhanced if it is aligned with the goals of other like-minded individuals and institutions at such a level.

The United Nations identified eight action areas for such a *Culture of Peace* (UN 2006). Engineering has the potential to make a significant contribution to each area, which are here stated in italics, followed by some possible engineering contributions:

1. *Fostering a culture of peace through education.* The task for engineering here is (at least) twofold. Firstly, to ensure that engineering education equips engineers to recognise and act on their professional social responsibility. Secondly, to educate the general public, and especially politicians and other decision-makers, about the contribution engineering can make to the wellbeing of all.
2. *Promoting sustainable economic and social development.* Here engineers are already making contributions in many ways and especially through physical infrastructure development. Particularly noteworthy is the contribution of NGOs such as Engineers Against Poverty and Water Aid. A major challenge for societies at all stages of development is the provision of sustainable energy sources at reasonable cost.
3. *Promoting respect for human rights.* Individual engineers and engineering enterprises should give full recognition to national and international law before undertaking activities and should review current activities in the framework of such law. It would be beneficial if the engineering profession adopted a principle of respect for individual persons aligned with that shown by the medical profession.
4. *Ensuring equality between men and women.* The engineering profession needs to promote a much greater participation of women in its work, as is already happening in the case of medical engineering. Sociological studies of ethical priorities suggest that the presence of a significant number of women engineers would in itself help to promote a culture of peace and care in the profession.
5. *Fostering democratic participation.* Engineering can contribute by providing appropriate technologies for the effective distribution of information allowing informed choice. Engineering can also contribute to the detection and prevention of organised crime and terrorism.
6. *Advancing understanding, tolerance and solidarity.* UNESCO has particularly identified the role of cultural heritage projects in promoting reconciliation. Many of these projects, such as the restoration of ancient buildings and artefacts, benefit from advanced engineering techniques. The reconstruction of the historic bridge at Mostar in Bosnia-Hercegovina is an iconic example of such a role for engineering.

7. *Supporting participatory communication and the free flow of information and knowledge.* Engineering has a key role to play in ensuring that digital technologies are widely available in suitable forms at reasonable cost. A specific challenge is to ensure that the development of the poorest countries can also benefit from modern information technologies.
8. *Promoting international peace and security.* The greatest challenge is to ensure that politicians and other decision-makers realise the benefits of engineering for the promotion of peace and security defined in the broadest possible terms and, in particular, that they appreciate the role of engineering in the non-military prevention and resolution of conflict.

In some cases these engineering contributions are already being carried out to a very great extent. Others are innovative. It is pertinent to note that even those actions that appear societal can benefit from appropriate engineering. For example, drilling convenient wells can promote gender equality as women are freed from the often onerous task of collecting water from a remote source. Most importantly, a real benefit of considering these action areas together is to use the context of an authoritative set of priorities to provide a real focus for the future tasks of individual engineers, engineering institutions and engineering enterprises.

A key responsibility of any society is to ensure the security of its citizens. The role of engineering in contributing to such security has most usually been considered as the development, manufacture and use of military equipment so as to ensure dominance if tensions result in violence. War has been, and remains, the normal business of engineering, and the present book has been critical of the extensive involvement of engineers in such activities. However, the promotion of a *Culture of Peace* suggests more effective ways for individual engineers and engineering enterprises to contribute to international security. Indeed, the goals of such a culture are supported by perceptive analyses of current threats to peace, and of the most effective responses, provided by independent expert organisations such as the Oxford Research Group (ORG 2006). The Group identifies four factors as the root causes of conflict and insecurity: climate change, competition over resources, marginalisation of the majority world and global militarisation. The Group characterises the predominant current responses as a *control paradigm* – an attempt to maintain the existing state of affairs through military means. They propose that a more effective approach is a *sustainable security paradigm* – to cooperatively resolve the root causes of these threats using the most effective means available.

It will be noted that engineers can play a major role in resolving each of the four root causes identified. For example, development of renewable energy sources can reduce climate change; improved efficiency and recycling can reduce resource competition; generation of wealth can diminish marginalisation; and reducing or halting weapons development can limit militarisation.

Such creative analysis is also starting to become part of government policy in the UK. Despite the modest size of its population and its peaceful geographical location, the UK has the second highest military budget in the world in cash terms, and the fifth highest in purchasing power (after the US, China, India and Russia). UK government strategy on security therefore has global significance, and it has recently been clarified in a single document for the first time. This publication makes clear that, "The broad scope of this strategy also reflects our commitment to focus on the underlying drivers of security and insecurity, rather than just immediate threats and risks" (Cabinet Office 2008: 4). It further recognises that climate change, competition for energy and water stress are "the biggest potential drivers of the breakdown of the rules-based international system and the re-emergence of major inter-state conflict, as well as increasing regional tensions and instability" (Cabinet Office 2008: 19).

The consonance of these aspects of the strategy with the Oxford Research Group's analysis is striking, and the challenge to engineers is again clear. The UK government has also created an initiative specifically "to help manage conflict and stop it spilling over into violence … Preventing conflict is better and more cost effective than resolving it" (FCO 2003: 3, 5). However, though this strategy and initiative are very welcome, there is at present a substantial tentativeness about their implementation. Thus, the total UK budget for conflict prevention and peacekeeping has been estimated to be only about 2% of that for direct military expenditure, and of the same order as subsidies to arms exporters (Elsworthy 2004). Engineers can play a very strategic role in changing this tentativeness to genuine commitment especially through taking a greater responsibility for informing politicians and other decision makers about the capabilities of engineering for resolving the root causes of conflict: none of the sources cited in this section refers explicitly to engineering.

Achievement of the aspiration in engineering to give priority to helping people over technical ingenuity would be greatly facilitated if priority is also given to the promotion of a culture of peace *within* the profession. This will need the incorporation of increased degrees of compassion and generosity in the fulfilment of our tasks. Indeed, an apt summary of the outlook of the present book could be: *Promoting a culture of peace within engineering: engineering for the promotion of a Culture of Peace,* understanding peace in its most complete sense.

7.7 An Aspirational Engineering Ethical Ethos Across Cultures

The aspirational engineering ethic developed in the present book has been expressed using philosophical concepts developed in the context of a Western cultural background. However, engineering is carried out across national and cultural boundaries, so it is pertinent to investigate whether the analysis developed is compatible with ethical views from very different philosophical approaches and cultures. The priority of mutual care in different religious and philosophical traditions has been mentioned in Chap. 6. Systematic comparison is beyond the present scope, but it is useful to consider one very different cultural tradition as an example to explore the possibility of ethical understanding and agreement across conceptual schemes.

The Buddhist view of the world is conceptually very different to that of Western culture. Hence, it presents a challenging example for cross-cultural evaluation. There has recently been great interest in trying to compare European and Buddhist understandings of ethics. Such comparison needs to be undertaken with sensitivity so as to avoid forcing one of these different understandings into a pattern provided by the other, for important characteristic features may then be lost. However, in carrying out a cautious comparison it may be noted that Buddhism has rules the following of which is close to an absolute duty. Again, Buddhist scriptures also advise thorough reflection on the consequences of actions. Nevertheless, there appears to be a consensus that Buddhist ethics is closer to virtue ethics than to any other Western approach (Keown 1996: 25).

The quintessentially Buddhist virtue of *ahimsã*, a "deeply positive feeling of respect for all living things", has already been mentioned in Chap. 6. Such a feeling is associated in the West with concepts such as respect for life or the sanctity of life. However, these terms have most often been applied to human life alone. The Western views that are closest to *ahimsã* are probably those associated with the deep ecology movement such as the "integral value" of all living things. Another virtue which is important for lay Buddhists in particular is generosity at all levels in society, between family members, friends and also strangers. Compassion is also important, and is one of the states of mind that is cultivated through the practice of meditation. These and other virtues may be regarded as being derived from the "cardinal virtues" of non-attachment, benevolence and understanding (Keown 1996: 12–13). It is clear that these Buddhist virtues are closely related to those of Western traditions, though the inclusion of non-attachment suggests a more spiritual starting point than has been common in the West, at least in recent times.

One of the most important modern developments in Buddhism has been the emergence of "Engaged Buddhism". This emergence seems to have been stimulated at least in part by increasing contacts between East and West. Engaged Buddhism is especially concerned with social justice. The main characteristics of socially engaged Buddhism have been summarised as follows (Queen 2000: 6–7):

Awareness Mindfulness or alertness to the situation around us.
Identification Empathy for others, sympathy, even as far as a co-feeling or fellow-feeling that dissolves the boundary between oneself and the other.
Action "Once there is seeing, there must be action".

Further, most engaged Buddhists view their action as having three characteristics:

non-violent	observant of the precept not to harm others
non-hierarchical	believing in the equal dignity of all persons
non-heroic	believing that effective social change requires collective, "grass-roots" activity

There are clearly close parallels between the analysis developed in the present book and the approach of Engaged Buddhism. The Buddhist characteristics of *awareness* and *identification* are analogous to those expressed by Buber as *I-Thou* and by Levinas as *the face* (see Chap. 6). Indeed, the Buddhist identification is demanding in so strong a way as to allow comparison with that described by Levinas as substitution. The principles of *action*, that it should be *non-violent*, *non-hierarchical* and *non-heroic* are also closely analogous to the aspirational ethical ethos that has been proposed for engineers. Furthermore, Buddhism has a view of virtuous living analogous to that of a *narrative unity*, "Buddhism is first and foremost a path of self-transformation … the cultivation of particular virtues (paradigmatically wisdom and compassion) leading step by step to the goal of complete self-realization known as nirvana" (Keown 1996: 25). However, there are also important differences. For example, the Western tradition does not have the emphasis on the "oneness" of all things as conceived in Buddhism and the mention of *nirvana* provides a reminder that there are great underlying conceptual differences between Eastern and Western viewpoints. Nevertheless, the aspirational ethos proposed in the present book has many parallels with the very different Buddhist view of the world, certainly enough to allow discussion and even consensus transcending the cultural differences.

7.8 Final Suggestions

Some readers may have worked diligently through the text to reach this point. Others may have jumped here to find the conclusions. To both categories, and to those with intermediate reading patterns, I offer the following final suggestions:

1. We need to restore to engineering a balanced prioritisation of helping people over technical ingenuity. In particular, we need to be aware that technological artefacts are merely goods of engineering, a means toward achieving the end of the promotion of human flourishing through contribution to material wellbeing.
2. Engineers should seek to promote the wellbeing of all people irrespective of national boundaries. The term "all" should be understood as referring both to human collectivity and to the human quality in each individual.
3. Each and every engineer needs to take an active role in considering the ethical implications of his or her work. It is no longer credible to be an engineer who "does what he is told but does not know what he is doing".

4. The Good Samaritan was the person who approached the wounded traveller and made him his neighbour. Likewise, our aspiration as engineers should be to hear the voice in need which cries, “It’s me here!” and to reply, “Here I am, how can I help you?”.

References

Borges JL (1998) Emma Zunz. In: Collected fictions, Hurley A (ed), Penguin Books, New York. Originally published in El Aleph, 1949

Cabinet Office (2008) The national security strategy of the United Kingdom, TSO, London

Elworthy, S (2004) Cutting the costs of war, ORG, Oxford

Engineers Against Poverty (2008) http://www.engineersagainstpoverty.org

Foreign and Commonwealth Office (2003) The global conflict prevention pool, FCO, London

Golding W (1964) The spire, Faber and Faber, London

Golding W (1988) A moving target, Faber and Faber, London

Hoare M (2007) Letter sent to potential students

Institution of Civil Engineers (2008) http://www.ice.org.uk

Keown D (1996) Buddhism: a very short introduction, Oxford University Press, Oxford

Lehrer T (1965) Wernher von Braun, That was the year that was, Reprise/Warner Brothers Records

McGonagall W (2004) The Tay Bridge disaster, In: McGonagall W: a selection, Walker CSK (ed), Birlinn, Edinburgh. First published 1890

Naess A (1989) Ecology, community and lifestyle, Cambridge University Press, Cambridge. Originally published as Økologi, samfunn og livsstil, Universitetsforlaget, Oslo, 1976

Orwell G (2000) Politics and the English language, in “Essays”, new edition, Penguin Books, London. Originally published 1946

Oxford Research Group (2006) Global responses to global threats, ORG, Oxford

Queen CS (2000) Introduction: a new Buddhism, in “Engaged Buddhism in the West”, Queen CS (ed), p. 1–31

Ricoeur P (1995) Figuring the sacred, Augsburg Fortress, Philadelphia. Chapter originally published 1991

Royal Academy of Engineering (2007) Educating engineers for the 21st century, RAE, London

Royal Academy of Engineering (2008) An engineering ethics curriculum map, RAE, London

Scientists for Social Responsibility (2008) http://www.sgr.org.uk

United Nations (1999) Declaration and programme of action on a culture of peace, General Assembly, A/RES/53/243, UN, New York

United Nations (2006) Culture of peace, General Assembly, A/61/175, UN, New York

Vesilind PA (2005) The evolution of peace engineering, in “Peace Engineering”, Vesilind PA (ed) Lakeshore Press, Woodsville, p. 1–11

Williams B (1996) History, morality and the test of reflection, in “The sources of morality”, Korsgaard CM, Cambridge University Press, Cambridge, 210–218

World Federation of Engineering Organisations (2001) The WEFO model code of ethics, WFEO, Paris

Index

Lightning Source UK Ltd.
Milton Keynes UK
UKOW052154021012

199959UK00010B/21/P